Analysing Qualitative Data

A. E. Maxwell M.A., Ph.D.

Senior Lecturer in Statistics
The Institute of Psychiatry
University of London

LONDON

CHAPMAN AND HALL

First published 1961
by Methuen & Co. Ltd
Reprinted 1964, 1967
Reprint 1971 published by
Chapman and Hall Ltd
11 New Fetter Lane, London EC4P 4EE
Reprinted 1975 as a Science Paperback

© *1961 A. E. Maxwell*

Printed and bound in Great Britain by
Redwood Burn Limited, Trowbridge & Esher

ISBN 0 412 21300 1

Distributed in the U.S.A.
by Halsted Press, a Division
of John Wiley & Sons, Inc., New York

Library of Congress Catalog Card Number 75-10907

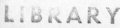

Contents

ANALYSING QUALITATIVE DATA

General Editor's Preface

It is not so very long ago that up-to-date text-books on statistics were almost non-existent. In the last few decades this deficiency has largely been remedied, but in order to cope with a broad and rapidly expanding subject many of these books have been fairly big and expensive. The success of Methuen's existing series of monographs in physics or in biology, for example, stresses the value of short inexpensive treatments to which a student can turn for an introduction to, or a revision of, specialized topics.

In this new Methuen series the still-growing importance of probability theory in its applied aspects has been recognized by coupling together Probability and Statistics; and included in the series are some of the newer applications of probability theory to stochastic models in various fields, storage and service problems, 'Monte Carlo' techniques, etc., as well as monographs on particular statistical topics.

Author's Preface

This book, which might have been given the title χ^2 *Tests*, aims at providing the research worker with a simple but up-to-date account of statistical techniques available for the analysis of qualitative data. It is complementary to *Experimental Design in Psychology and the Medical Sciences* in which continuous variables only are considered.

While in preparation the manuscript was read by Dr D. R. Cox. His penetrating and scholarly criticisms were greatly appreciated and I am much indebted to him. I am indebted too to Professor M. S. Bartlett for valuable editorial comments, and to Professor W. G. Cochran from whose work I borrowed freely.

The illustrative examples were drawn largely from studies by my psychiatric and psychological colleagues at the Institute of Psychiatry, but as I have occasionally bent their data to fit my own requirements it is best not to mention them individually.

Finally, I must thank Professor E. S. Pearson, on behalf of the Biometrika trustees, for permission to reproduce in abridged form Table 8 from *Biometrika Tables for Statisticians*, Vol. I, published by the Cambridge University Press.

CHAPTER I

The Chi-Square Test

Introduction. This book commences with a discussion of the chi-square (χ^2) test. It will be introduced in this and the next chapter, while in the following three chapters important recent developments of the test—which are still not widely known or employed—will be reviewed and demonstrated.

The chi-square test is often used in experimental work where the data consist in frequencies or 'counts'—for example the number of boys and the number of girls in a class that have had their tonsils out—as distinct from quantitative data obtained from the measurement of continuous variables such as height, temperature, and so on. The most common use of the test is probably as one of *association*, and recently a lot has been heard about it in connection with investigations concerning smoking and carcinoma of the lung. The test has also been used extensively in experiments designed to assess the effect of inoculation in immunising people against disease, and in clinical trials involving the use of drugs. In this book many examples of its use in investigations in psychology, psychiatry and social medicine will be given. These sciences are as yet in the early stages of development and studies in them are still characterised by a search for the variables basic to them. This search is often one for general relationships and associations—however amorphous they may appear at first—between the phenomena being studied, and here the chi-square test is often helpful.

As previously stated the test is frequently employed as one of *association*, and when the word 'association' is used in a statistical sense a comparison is implied. For instance if we say that there is an 'association' between inoculation and immunisation against some disease we mean that the proportion of inoculated people who contract the disease is different from the proportion of uninoculated people who do so. Of course the two proportions might be expected to differ in some measure due solely to chance factors of sampling, and for other reasons which might be attributed to random causes, but the test enables us to calculate the probability that a difference as great or greater than that obtained could have arisen in this way.

Classification. In introducing the test the first matter requiring attention is that of classification. Normally it is possible to classify the members of a *population*—a generic term denoting any well-defined class of people or things—in many different ways. People, for instance, may be classified into male and female; into married and single; into those who are eligible to vote and those who are not; and so on. These are examples of dichotomous classifications. But multiple classifications also are common as when people are classified into left-handed, ambidextrous, right-handed; or, for the purposes of a Gallup poll, into those who intend to vote *a* Conservative, *b* Labour, *c* Liberal, *d* those who have not yet made up their minds, and *e* others. Whether the classification is dichotomous or multiple it must be exhaustive, and the categories into which it divides the members of the population must be mutually exclusive. A classification is exhaustive when it provides sufficient categories to accommodate all members of the population. The categories are mutually exclusive when they are so defined that each member of the population can be correctly allotted to one, and only one category. At first sight it might appear that the requirement that a classification be exhaustive is very restrictive. We might, for example, be interested in carrying out a Gallup poll, not on the voting intentions of the electorate as a whole, but only on those of university students. The difficulty is resolved if the definition of a population is recalled. The statistical definition of the word is more fluid than its definition in ordinary usage, so that it is quite in order to define the population in question as 'all university students eligible to vote'. Categories too are adjustable and may often be altered or combined; for instance, in the voting example it is unlikely that much information would be lost by amalgamating categories *d* and *e*.

Contingency tables. When the members of a sample have been doubly classified, that is classified in two separate ways, the results can be arranged in a rectangular table. For instance the number of people on record who, in 1956, died of tuberculosis in England and Wales was 5375. Of these 3804 were males and 1571 were females: 3534 males and 1319 females died of tuberculosis of the respiratory system, while the remainder died of other forms of tuberculosis. These data can be tabulated as follows:

TABLE 1.1

Deaths from Tuberculosis in England and Wales in 1956

	Males	Females	Total
Tuberculosis of respiratory system .	3534	1319	4853
Other forms of tuberculosis . .	270	252	522
Tuberculosis (all forms) . . .	3804	1571	5375

A table such as this is known as a *contingency* table, and this 2×2 example (the members of the sample having been dichotomised in two different ways) is its simplest form. Had the classifications been multiple rather than dichotomous the table, while still being square or rectangular, would have had many more cells.

The entries in the cells of a contingency table may be frequencies, as in Table 1.1, or the frequencies may be transformed into proportions or percentages. It is important, however, to note that in whatever form they are presented the data were frequencies or 'counts'—rather than continuous measurements—in the first place, otherwise the chi-square test could not be applied to them. In short the chi-square test can be used only on discrete data: for the purposes of the test, of course, continuous data can often be put into discrete form by the use of intervals on a continuous scale. Age, for instance, is a continuous variable, but if people are classified into different age groups then the intervals of time corresponding to these groups can be treated as if they were discrete units.

Samples. In Table 1.1 all the known deaths from tuberculosis in England and Wales for the year 1956 are recorded. But in statistical investigations in general it is seldom the case that information for the whole population is available. Only a sample from it is used. Indeed one main function of statistical science is to demonstrate how valid inferences about some population may be made from an examination of the information supplied by a sample. An essential step in this process is to ensure that the sample taken is an unbiased one. This can be achieved by drawing what is called a *random* sample, that is one in which each member of the population in question has an equal chance of being included. Alternatively we might try to draw a sample that was representative of the population. In the case of deaths from tuberculosis we might, at the beginning of a given year, divide the members of the population into groups according to socio-economic status, choose a random sample from each group proportional in size to the number of people in that socio-economic group, and then at the end of the year ascertain the number in that group that had died from tuberculosis. Sampling done in this way is known as stratified sampling.

Sampling, too, can be done along the time continuum; in this sense the data in Table 1.1—though representing a complete year-group of deaths from tuberculosis—may be thought of as a sample. Suppose we desire to compare the proportion of males who die from a particular form of tuberculosis with the proportion of females who die from the same cause. If it can safely be assumed that these proportions are stable over a period of years then a test of the required difference can be made without bias on the data for any one year within the period provided the year is chosen randomly. In the present instance it will be convenient to think of the year 1956 as having been chosen in this manner.

Independent classifications. When a sample of people or things have been classified according to two attributes, as in Table 1.1, it is reasonable to enquire whether the two classifications are independent. To answer this question it is necessary to get clear just what independence between the classifications would entail. For this purpose it will be helpful to rearrange Table 1.1 leaving the marginal frequencies as they are but replacing the frequencies in the cells of the body of the table by the letters E_1 to E_4. The adjusted table now becomes:

TABLE 1.2

Table 1.1 with Cell Frequencies Replaced by Letters

	Males	Females	Totals
Tuberculosis of the respiratory system	E_1	E_2	4853
Other forms of tuberculosis . .	E_3	E_4	522
Tuberculosis (all forms) . . .	3804	1571	5375

Looking at the column of marginal totals on the right of the table it is seen that the proportion of deaths, males and females combined, due to tuberculosis of the respiratory system, is

$$4853/5375 = 0.903$$

Now if the form of tuberculosis from which people die is independent of sex—which is another way of saying that the two classifications are independent—we would expect that the proportion of males that died from tuberculosis of the respiratory system would equal the proportion of females that died from the same cause, and consequently would equal the proportion of the total group, 0.903, that died from this cause. The expected values E_1 and E_2 must then be chosen so that both $E_1/3804$ and $E_2/1571$ are equal to 0.903. Taking the data direct from Table 1.2 the equations for E_1 and E_2 are:

$$E_1/3804 = 4853/5375 \qquad \ldots 1.1$$

and

$$E_2/1571 = 4853/5375 \qquad \ldots 1.2$$

On multiplying each side of equation 1.1 by 3804 an equation giving the value of E_1 explicitly is obtained. It is:

$$E_1 = (4853 \times 3804)/5375 = 3434.6 \qquad \ldots 1.3$$

This value of E_1 is what one would expect to find in the top left-hand cell of Table 1.1 were the forms of tuberculosis from which people die independent of sex, for then the proportion of males—$3434.6/3804 = 0.903$—that died from tuberculosis of the respiratory system would be equal to the proportion of the total group that died from the same cause.

On substituting the value of E_1 obtained from equation 1.3 in

Table 1.2 the values of E_2, E_3 and E_4 can be deduced, for the sum of E_1 and E_2 must equal 4853, etc. Moreover, the value of E_2 found in this way can be checked by employing equation 1.2. Similarly the values of E_3 and E_4 can be checked by setting up equations for them similar to equations 1.1 and 1.2. The equation for E_4, for instance, is:

$$E_4/1571 = 522/5375 \qquad \ldots 1.4$$

which gives $E_4 = (522 \times 1571)/5375 = 152 \cdot 6$.

Replacing the E's in Table 1.2 by their calculated values the table of expected values is obtained. These are the values that one would expect to find in the cells in the body of Table 1.1 were the two methods of classification independent.

TABLE 1.3

Expected Frequencies on the Assumption of Independent Classifications

	Males	Females	Total
Tuberculosis of the respiratory system	3434·6	1418·4	4853
Other forms of tuberculosis . .	369·4	152·6	522
Tuberculosis (all forms) . . .	3804·0	1571·0	5375

Though fractions do not appear in contingency tables showing observed frequencies they may appear in tables of expected frequencies and should—especially when the sample size is small—be retained, as they increase the accuracy of subsequent calculations.

Calculating chi square. If we refer again to the data we see that the observed frequencies, Table 1.1, differ considerably from the frequencies to be expected, Table 1.3, were the two classifications of the people in the sample independent of each other. The question now is whether the discrepancies between the entries in the two tables are such as could have arisen from random sampling errors alone or whether they are indicative of a real difference between males and females. It is this question that the chi-square test helps us to answer.

In introducing the test it is necessary to digress a little to say a word or two about statistical distributions. The best known is the normal distribution, and the author has given a simple account of it elsewhere (Maxwell, 1958). It is a symmetrical distribution and the mathematical formula for it enables us to calculate the probability of occurrence of deviations, as great or greater than a given amount, from its central value. Apart from the normal distribution the χ^2 distribution is perhaps the next best known. It is the probability distribution of the sum of the squares of a number of independent variates which are normally, or approximately normally, distributed with means of zero and standard deviations of unity. The probability distribution depends upon the number of independent variates, or more strictly upon the number of degrees of freedom (discussed in the next section), associated with the variates.

The sum of squares in question is generally denoted by χ^2 and when dealing with frequency data, that is with 'counts', it can be shown that—provided the expected frequencies (E_i) are not too small—the expression:

$$\sum \frac{(O_i - E_i)^2}{E_i} \qquad \ldots 1.5$$

is distributed approximately as χ^2. In this expression O_i stands for the observed frequencies and E_i for the expected frequencies and i runs from 1 to n, the number of cells in the contingency table.

The mathematical expression for the χ^2 distribution need not be discussed here for a table (Appendix, Table A) is available from which—on the null hypothesis that the observed and the expected frequencies do not differ, and knowing the degrees of freedom—the probability P of exceeding any particular value of the expression 1.5 can be read off. To guard against bias in the test, should one or more of the expected frequencies be small, a *correction for continuity* should be made when using the expression 1·5. This will be introduced shortly.

In the case of large samples where all the expected frequencies are greater than say 5, formula 1.5 states that to evaluate χ^2 the difference between the observed and the expected frequencies for each cell of the contingency table is found; this difference is then squared and the answer divided by the expected frequency concerned. The sum of the results of these calculations for all the cells is the required value of χ^2.

To perform the calculations it is useful to arrange the observed and the expected frequencies as shown in Table 1.4. The differences $(O_i - E_i)$ are then obtained, a check being that they add to zero. Next the differences are squared and each squared difference is divided by the expected frequency in its own row. These values appear in column five of the table and their sum gives the required value of x^2.

TABLE 1.4

Calculating x^2 for the Data in Table 1.1

O	E	$(O-E)$	$(O-E)^2$	$(O-E)^2/E$
3534	3434·6	99·4	9880·36	2·88
1319	1418·4	−99·4	9880·36	6·97
270	369·4	−99·4	9880·36	26·75
252	152·6	99·4	9880·36	64·75
5375	5375·0	0·0		$x^2 = 101·35$

Testing the significance of an obtained value of x^2. To assess the significance of the value $x^2 = 101·35$ it is referred to the chi-square table, Appendix A. The numbers in the first column of that table, under the letter 'n', are known as the *degrees of freedom*. These can be explained by referring to Table 1.2. There it was seen that once the value of any one of the E's had been determined all the other E's could be deduced. In other words, when the marginal totals of a 2×2 table are given, only one cell in the body of the table can be filled arbitrarily. This fact is expressed by saying that a 2×2 table has only one degree of freedom. Had the classifications in the example been multiple, yielding a contingency table with, say, r rows and c columns then the degrees of freedom associated with it—as shall be seen later—would have been $(r-1)(c-1)$. Indeed if in this expression r and c are both put equal to 2 the value 1, as in the present example, is obtained.

Since Table 1.1 has only one degree of freedom we enter the x^2 table, Appendix A, with n equal to one, that is we look along the first row of the table. The largest value in that row is 10·828 under the probability (P) level 0·001. A value of chi square equal to or greater than 10·828 would be expected to occur by chance only once in a thousand times if the null hypothesis is true. Since our value of chi square of $101 \cdot 35$ is much greater than $10 \cdot 828$ it would be expected to occur even less frequently still. This result can hardly be attributed to chance, so we conclude that the two classifications of the people in our sample are not independent. Put in another way, the evidence shows that the proportion of males that died from tuberculosis of the respiratory system, namely $3534/3804 = 0 \cdot 929$, is significantly different from the proportion of females, namely $1319/1571 = 0 \cdot 840$, that died from the same cause. The chi-square test of association is seen then to be a test of whether certain proportions differ. This interpretation is a useful one to keep in mind as it assists in the interpretation of the results.

In carrying out a test of significance it is customary to state the hypothesis being tested in formal terms. For the example under consideration the *null hypothesis* that the two classifications are independent is set up. The value of χ^2 is computed and the number of degrees of freedom on which it is based ascertained. This value is then referred to the table of values of χ^2 and the probability of a value as great or greater than that obtained is found. If this probability is small it indicates that the result obtained would be expected to occur by chance very rarely and so is indicative of a real difference between the proportions being compared. Critical probability levels in common use are the values $0 \cdot 05$ (the 5 per cent level) and $0 \cdot 01$ (the 1 per cent level).

Interpretation. When the probability level obtained from a chi-square test is so small that it is decided to discard the null hypothesis then an association between the two methods of classification of the members of the population concerned is claimed. Put in different words the two methods of classification are not independent, and it is natural at this stage to enquire why this is so. Interpretation of the result requires careful handling for questions of association are frequently entangled with questions of 'cause' and 'effect'—which proverbially are enigmatic. For this reason it is necessary to stress that the establishment of a statistical association by means of the chi-square test does not necessarily imply any causal relationship between the attributes being compared, but it does indicate that the reason for the association is worth investigating.

In the example given earlier—where the data are used primarily to demonstrate a statistical technique—though a satisfactory explanation of the association found cannot be offered, it is relevant to note that (i) respiratory tuberculosis as a cause of death for people under 25 years has virtually disappeared in Britain, (ii) whereas the death rate from this disease is steadily falling for people in middle age the rate of fall for males is slightly greater than that for females so there is a tendency for sex differences to disappear.

Simplifying the calculations. Having acquired some idea about the argument behind the chi-square test it will be well to see how the calculations can be simplified. For this purpose let us write the fourfold table, using letters instead of numbers, as below:

TABLE 1.5

The General 2 × 2 Contingency Table

		Not-A	A	Total
B	. .	a	b	a+b
Not-B	. .	c	d	c+d
Total	. .	a+c	b+d	N

where a sample of size $N = (a+b+c+d)$ has been classified in two ways according to (1) whether they do or do not possess the attribute A, and (2) whether they do or do not possess the attribute B. Using these symbols it can be shown that χ^2 is given by the formula:

$$\chi^2 = \frac{N(ad-bc)^2}{(a+b)(c+d)(a+c)(b+d)} \qquad \ldots 1.6$$

with one degree of freedom.

Applying this formula to the data in Table 1.1 we get:

$$\chi^2 = \frac{5375(3534 \times 252 - 1319 \times 270)^2}{4853 \times 522 \times 3804 \times 1571}$$

$$= 101 \cdot 35$$

which agrees with our earlier calculations.

Small Expected Frequencies

As was stressed earlier the mathematical derivation of the distribution:

$$\chi^2 - \sum \frac{(O-E)^2}{E}$$

from which the tabulated values (Appendix, Table A) are derived, assumes that the expected values (E) are large; indeed the proof is valid only on the assumption that the expected values tend to infinity. On this account most writers warn against using the test when any of the E-values are less than 5. This warning means that the use of the test is restricted to large samples; but fortunately where small samples are concerned the difficulty can be overcome in the case of the 2×2 table in a number of ways. One way is to apply a *correction for continuity*, known as the *Yates correction* (Yates, 1934). This is done by subtracting $\frac{1}{2}$ from the positive discrepancies ($O-E$) and by adding $\frac{1}{2}$ to the negative discrepancies before these values are squared (see Table 1.4). Alternatively, the correction can be incorporated in equation 1.6, which then reads:

$$\chi^2 = \frac{N(|ad-bc| - 0\cdot5N)^2}{(a+b)(c+d)(a+c)(b+d)}, \quad \text{d.f.} = 1 \qquad \ldots 1.7$$

In this equation the vertical lines either side of the expression ($ad-bc$) mean that the absolute value of this expression is always taken, that is we take its value as positive whether it turns out to be positive or negative.

It is desirable to apply the Yates correction at all times, whether or not the E-values are greater than 5, but it is essential to do so in cases like that discussed below in which the sample size is small.

An example in which the sample is small. In a broad general sense mental patients can be classified as psychotics or neurotics. A psychiatrist, while studying the symptoms of a random sample of 20 from each of these populations found that whereas 6 patients in the neurotic group had suicidal feelings, only two in the psychotic group suffered in this way and he wished to test if there is an association between the two psychiatric groups and the presence or absence of suicidal feelings.

TABLE 1.6

The Incidence of 'Suicidal Feelings' in Samples of Psychotic and Neurotic Patients

	Psychotics	Neurotics	Total
Suicidal Feelings . .	2	6	8
No Suicidal Feelings.	18	14	32
Total	20	20	40

Following the same procedure as in our earlier example the frequencies which we would expect to find in the cells of the body of the table, under the hypothesis of independence of the two methods of classification, can be calculated. The expected frequency (E_1) for the top left-hand cell is given by the equation:

$$E_1/20 = 8/40$$

so that E_1 is equal to $(8 \times 20)/40$, namely 4. The complete table of expected frequencies, for the given marginal frequencies, can now be deduced. It is:

TABLE 1.7

Expected Frequencies for the Data in Table 1.6

Psychotics	Neurotics	Total
4	4	8
16	16	32
20	20	40

Here two of the expected frequencies fall below 5, so one would naturally wish to employ the Yates correction when calculating the value of χ^2. Following the procedure of Table 1.4 the details of the calculations are as given in Table 1.8.

TABLE 1.8

Calculating χ^2 for the Data in Table 1.6 Using the Yates Correction

(1)	(2)	(3)	(4)	(5)	(6)
		Discrepancy	Corrected		
O	E	(O − E)	discrepancy	(col. 4)^2	(col. 4)^2/(col. 2)
2	4	−2	−1·5	2·25	0·562500
6	4	2	1·5	2·25	0·562500
18	16	2	1·5	2·25	0·140625
14	16	−2	−1·5	2·25	0·140625
					1·406250

The value of χ^2 is $1 \cdot 40625$ which with one degree of freedom is not significant at the 5 per cent level. Indeed if this value is referred to a detailed table of χ^2 for one degree of freedom (Kendall, 1952, Appendix, Table 6) it is found to correspond to a probability $P = 0 \cdot 23572$, so that discrepancies as large as those obtained might be expected to occur by chance about one in five times even with methods of classification which were independent.

The value of χ^2 just computed could also have been found by equation 1.7, which involves less computation. The details are as follows:

$$\chi^2 = \frac{40(|28 - 108| - 20)^2}{8 \times 32 \times 20 \times 20}$$

$$= \frac{40(80 - 20)^2}{102400} = 1 \cdot 40625, \text{ as before.}$$

In this example had the Yates correction not been employed the value obtained for χ^2 would have been $2 \cdot 5$, which corresponds to a probability of $P = 0 \cdot 11385$ and differs considerably from the value obtained above.

Fisher's Exact Test for 2×2 Tables

An exact method for testing the hypothesis of independence in 2×2 contingency tables will now be described. Though a bit laborious it is the only safe method to employ when the sample size is less than about 40 and one or more of the expected frequencies falls below 5. The test will be illustrated on the data in Table 1.6 and the result will furnish a check on the effectiveness of the Yates correction used above.

Under the assumption of independence between the two classifications in a 2×2 table the exact probability (P) of a particular set of values of a, b, c and d occurring (Table 1.5), when the marginal frequencies are taken as fixed, is given by the equation:

$$p = \frac{(a+b)!\,(c+d)!\,(a+c)!\,(b+d)!}{a!\,b!\,c!\,d!\,N!} \qquad \ldots 1.8$$

where $a!$—read a-factorial—is the shorthand method of writing the product of a and all whole numbers less than it down to unity. For example:

$$5! = 5 \times 4 \times 3 \times 2 \times 1 = 120$$

By definition the value of $0!$ is unity. Luckily the values of the factorials have not to be worked out as they are given in Barlow's (1952) and in Fisher and Yates's tables (1957). The calculations too can be simplified by cancelling: a little thought will show that $20!/18!$, for example, reduces to 20×19.

Substituting the values in Table 1.6 in equation 1.8 we get:

$$p_2 = \frac{8! \times 32! \times 20! \times 20!}{2! \times 6! \times 18! \times 14! \times 40!} = 0 \cdot 095760 \qquad \ldots 1.9$$

The subscript to p is the smallest of the quantities a, b, c and d. In the present example the smallest entry is in the top left-hand cell of the table and is 2.

Returning to Table 1.6, and keeping in mind that the marginal frequencies are to be taken as fixed, the frequencies in the body of the table can be arranged in two other ways both of which would represent, had they been observed, more extreme discrepancies between the psychotic and the neurotic groups. These arrangements are:

TABLE 1.9

More Extreme Cell Frequencies than Those Observed—Table 1.6

(a)		Total		(b)		Total
1	7	8		0	8	8
19	13	32		20	12	32
20	20	40		20	20	40

Substituting in turn the values in these tables in equation 1.8 we get:

$$p_1 = 0 \cdot 020160 \quad \text{for Table 1.9a}$$

and

$$p_0 = 0 \cdot 001638 \quad \text{for Table 1.9b}$$

The probability P of obtaining the cell frequencies shown in Table 1.6, or frequencies more extreme than them, Tables 1.9a and 1.9b, is therefore the sum of the three probabilities:

$$p_2 = 0 \cdot 095760$$
$$p_1 = 0 \cdot 020160$$
$$p_0 = 0 \cdot 001638$$

$$P = 0 \cdot 117558$$

This sum is the probability that two or less of the psychotics show suicidal feelings. However, when performing the χ^2 test on the data in Table 1.6 the fact that the smallest entry in the contingency table happened to occur for psychotics rather than for neurotics was not taken into account so that if the result of this exact test is to be compared with the result of the χ^2 test the probability $P = 0 \cdot 11756$ should be doubled: this is legitimate when row or column sums are equal. The probability now is $0 \cdot 23512$ and since the probability, using the χ^2 test with the Yates correction was found to be $0 \cdot 23572$ the efficacy of the latter correction is clearly demonstrated.

Simplifying the calculations. When using Fisher's exact test the calculations can be simplified by the use of logarithms. Equation 1.9 for example can be written in the form:

$$\log p_2 = (\log 8! + \log 32! + \log 20! + \log 20!)$$
$$- (\log 2! + \log 6! + \log 18! + \log 14! + \log 40!) \quad \ldots 1.10$$

Tables of the logarithms of factorials are easily constructed or can be found in standard books of statistical tables (Lindley and Miller, 1953; Fisher and Yates, 1957). Referring to the tables the values of the logarithms on the right-hand side of equation 1.10 can be read off. They are:

$$\log p_2 = (4 \cdot 6055 + 35 \cdot 4202 + 18 \cdot 3861 + 18 \cdot 3861)$$
$$- (0 \cdot 3010 + 2 \cdot 8573 + 15 \cdot 8063 + 10 \cdot 9404 + 47 \cdot 9116)$$
$$= -1 \cdot 0187 = \bar{2} \cdot 9813$$

Hence p_2 is the antilogarithm of $\bar{2} \cdot 9813$ which is $0 \cdot 09579$. Apart from errors due to rounding this value is the same as that obtained in equation 1.9

Another way to reduce the calculations is to use tables of binomial coefficients (Hald, 1952). To do so equation 1.8 must be written in the equivalent form:

$$p = \begin{pmatrix} a+b \\ a \end{pmatrix} \times \begin{pmatrix} c+d \\ c \end{pmatrix} \div \begin{pmatrix} N \\ a+c \end{pmatrix} \qquad \ldots 1.8a$$

where $\begin{pmatrix} a+b \\ a \end{pmatrix}$, also written $_{(a+b)}C_a$, stands for the number of combinations of $(a+b)$ things taking a at a time. To take an example, if a is 5 and b is 3 then $_{(a+b)}C_a$ is $_8C_3$, the value of which is

$$\frac{8 \times 7 \times 6}{3 \times 2 \times 1} = 56$$

The values of expressions such as those given in equation 1.8a can, as mentioned above, be found in the tables referred to.

The Comparison of Frequencies in Matched Samples

One-to-one matching is frequently used by research workers to increase the precision of a comparison. This matching is usually done on variables such as age, sex, weight, I.Q. and the like, information about which can be obtained easily. Two samples matched in a one-to-one way must be thought of as correlated samples and consequently are not independent. As a result the ordinary χ^2 test is not strictly applicable for assessing the difference between frequencies obtained with reference to these samples, for it would lead the experimenter to discard the null hypothesis less frequently at any given level of significance than it should be discarded. The appropriate test for comparing frequencies in matched samples is one due to McNemar (1955). As an introduction to it let us look at Table 1.10 in which the presence or absence of an attribute A for two matched Samples I and II is shown.

TABLE 1.10

Frequencies in Matched Samples

		Sample I	
		A absent	A present
Sample II	A present	a	b
	A absent	c	d

Since we are concerned with the differences between Samples I and II the entries in the N-E and the S-W cells of the table do not interest us, for the frequency b refers to matched pairs both of whom possess the attribute, while the frequency c refers to pairs both of whom do not possess the attribute. The comparison is thus confined to the frequencies a and d, the former representing the number of matched pairs who possess the attribute if they come from Sample I and do not possess it if they come from Sample II, while the latter represents matched pairs for whom the converse is the case.

On the null hypothesis that the two samples do not differ as regards the attribute we would expect a and d to be equal, or, to put it in another way the expected values for the two cells would each be $(a+d)/2$. Now if the observed frequencies a and d and their expected frequencies $(a+d)/2$ are substituted in the expression 1.6 and the result is reduced to its simplest form we arrive at the expression:

$$\chi^2 = \frac{(a-d)^2}{(a+d)}$$

If a correction for continuity is applied this expression becomes:

$$\chi^2 = \frac{(|a-d|-1)^2}{a+d} \qquad \ldots 1.11$$

with one degree of freedom. This is McNemar's formula for testing for an association in a 2×2 table when the samples are matched. The vertical lines either side of $(a-d)$ in the numerator of the expression direct us to take the absolute value of this difference, that is to take the difference as being positive whatever its sign turns out to be. To illustrate McNemar's test let us now consider an example.

An example. A psychiatrist wished to assess the effect of the symptom 'depersonalisation' on the prognosis of depressed patients. For this purpose a number (23 as it proved to be) of endogenous depressed patients, who were diagnosed as being 'depersonalised', were matched one-to-one for age, sex, duration of illness, and on certain personality variables, with 23 endogenous depressed patients who were diagnosed as not being 'depersonalised'. The numbers of pairs of patients from the two samples who, on discharge after a course of E.C.T. were diagnosed as 'recovered' or 'not recovered' are given in Table 1.11.

TABLE 1.11

Recovery of 23 Pairs of Depressed Patients

		Depersonalised patients		
		Not recovered	Recovered	Total
Patients not depersonalised	Recovered	5	14	19
	Not recovered	2	2	4
	Total	7	16	23

An exact test of the null hypothesis for these data, as we shall see in Chapter 9, is given by considering the first three terms in the binomial expansion $(\frac{1}{2} + \frac{1}{2})^7$, but for the present let us consider McNemar's test.

From the table we find that a is 5 and d is 2. When these values are substituted in equation 1.11 we obtain:

$$x^2 = \frac{(|5-2|-1)^2}{5+2} = \frac{2^2}{7} = 0 \cdot 57$$

With one degree of freedom this value does not reach an acceptable level of significance so we conclude that 'depersonalisation' is not associated with prognosis where endogenous depressed patients are concerned.

Guarding against Biased Comparisons

Earlier in this chapter the need to use random or representative samples as a safeguard against obtaining biased results in an investigation was stressed. Now that we have a few examples to which to refer some further discussion of the matter will be helpful.

An important advance in the development of statistical science was achieved when the advantages of *design* in experimentation were realised (Fisher, 1935). These advantages result from conducting an investigation in such a way that environmental effects and other possible disruptive factors, which might make interpretation of the results ambiguous, are kept under control. But in many investigations in social medicine and in survey work in general, where the data are often of a qualitative kind (and χ^2 tests are commonly required). planned experiments are difficult to arrange (Taylor and Knowelden, 1957, Chapter 4). One of the problems is that the occurrence of the phenomenon being studied may be infrequent so that the time available for the investigation permits a retrospective study only to be undertaken. With such studies it is generally difficult to get suitable control data and serious objections to the samples one might draw, because of limitations in the population being sampled, often arise. Berkson (1946) has drawn attention to this point where hospital populations are concerned, and has demonstrated that the subtle differential selection factors which operate in the referral of people to hospital are likely to bias the results of investigations based on samples from these populations. His main point can be illustrated best by an example.

Suppose an investigator wished to compare the incidence of tuberculosis of the lung in postmen and bus drivers. He might proceed by drawing two samples from the entrants to these occupations in a given month or year and do a follow-up study, with regular X-ray examinations, over a period of years to obtain the information he required. He would, of course, be aware of the possibility that people, by reason of their family histories or suspected predispositions to special ailments, might tend to choose one occupation rather than the other, and he might take steps to control for such possibilities and to eliminate other possible sources of bias. But suppose that, since time and the facilities at his disposal did not permit a prospective study to be carried out, he decided to obtain his samples by consulting the files of a large hospital and extracting for comparison all the postmen and bus drivers found there, the data obtained might not give a true picture. For instance it might be the case that bus drivers, by virtue of the special responsibility attached to their jobs, were more likely than postmen to be referred to hospital should tuberculosis be suspected. If this were so a biased comparison would clearly result.

A biased comparison would also result were it the case that bus drivers, say, were prone to be affected by multiple ailments such as bronchitis and tuberculosis, or carcinoma of the lung and bronchitis, or all three, for these ailments would be likely to aggravate each other so that a bus driver might be referred to hospital for bronchial treatment and then be found to have tuberculosis. Were this a common occurrence then a comparison of postmen and bus drivers as regards the incidence of tuberculosis, based on such hospital samples, would not give a true reflection of the incidence of the disease in these occupations in the community.

The relevance of the above discussion to the interpretation of results from investigations such as those reported in this chapter can now be examined. For instance if we turn to Table 1.6 it is clear that the chi-square test applied to the data in it yields an unbiased result only in so far as we can be sure that the hospital populations of Psychotics and Neurotics from which the samples are drawn are not affected by differential selection. In particular we would want to satisfy ourselves that 'suicidal feelings' did not play a primary part in the referral of Neurotics to hospital in the first place. If it did the results given by the chi-square test would be biased.

But it is well to add that Berkson (1946) notes certain conditions under which unbiased comparisons can be made between samples drawn from sources in which selective factors are known to operate. For instance, if the samples of postmen and bus drivers drawn from the hospital files are selected according to some other disease or characteristic unrelated to tuberculosis, say those who on entry to hospital were found to require dental treatment, then a comparison of the incidence of tuberculosis in these men would yield an unbiased result.

As a means of avoiding a biased comparison between samples drawn from a biased source it might too be thought that a one-to-one matching of subjects from the populations to be compared would overcome the difficulty. But clearly this could not act as a safeguard. In the example discussed earlier in the chapter, in which the effect of the symptom 'depersonalisation' on the prognosis of endogenous depressed patients was assessed, were it the case—which is unlikely— that 'depersonalisation' itself was a primary factor in causing depressed patients to come to hospital then the result of the comparison made in that investigation would be open to doubt.

Enough perhaps has been said to warn the beginner against biased comparisons which arise from faulty sampling, but in concluding the section a word of warning should be given here about bias of a different sort. It results from the pooling of data that are not homogeneous. Reminders against this type of bias will be given continuously in succeeding chapters: a classic example is the creation of a significant chi square by pooling together two 2×2 contingency tables comparing two medical treatments by means of the two survival proportions, one table relating to males and one to females, though each table separately showed no differential effect of the two treatments.

Rearranging the Data in 2×2 Tables

Data presented in a 2×2 table can sometimes be meaningfully rearranged so that two separate null hypotheses can be tested on them. An example will illustrate the point. Two random samples of adults of sizes 41 and 22 respectively, drawn from two Indian tribes A and B, were administered the Rorschach test. The tribes, though living adjacent to one another, had very different customs and ways of life and were of contrasting personalities; the question was whether these differences were reflected in their Rorschach records.

TABLE 1.12

Allocation of Individuals to their Correct Tribe

		Psychologist's allocation		
		B	A	Total
True	$\{A$	6	35	41
allocation	$\{B$	10	12	22
Total	. .	16	47	63

The records for the total of 63 individuals were well shuffled and given to a psychologist with experience in Rorschach interpretation. The differences between the two tribes were explained to him; he was told how many people from each were in the combined sample and he was asked to allot the individuals to one or other of the tribes on the basis of his examination of the Rorschach records. The results are given in Table 1.12.

The first null hypothesis to be tested on the data was that there was no association between the psychologist's allocation and the true allocation of the members of the combined sample. Chi square, using the Yates correction, was found to be 5·644 which is significant at the 0·02 level, so the null hypothesis had to be discarded. In other words the psychologist's allocation of the people to their correct tribe on the basis of his analysis of their Rorschach records was better than chance. Arithmetically speaking the proportion 35/47 has been shown to be greater than the proportion 6/16.

Next the question arose as to whether the proportion of individuals correctly classified was the same for each tribe. These proportions are obtained from the diagonal cells of the above table and are $35/41 = 0·854$ for sample A and $10/22 = 0·455$ for sample B. To test whether these two proportions differ the data in Table 1.12 were rearranged as in Table 1.13.

TABLE 1.13

Data in Table 1.12 Rearranged According to Correct and Incorrect Classification

	Incorrect	Correct	Total
Sample A . .	6	35	41
Sample B . .	12	10	22
Total . .	18	45	63

For the data in the latter table chi square was found to be 9·305 which is significant beyond the 0·01 level of significance, consequently the null hypothesis that the proportion of individuals correctly classified for the two tribes did not differ had to be discarded. The psychologist had proved more efficient at classifying correctly individuals from tribe A than from tribe B after an examination of their Rorschach records.

Confidence Intervals and Relative Risks

So far in this chapter we have confined our attention to tests of significance of the null hypothesis (that no association exists) in 2×2 contingency tables. But important questions of *estimation* and of the setting up of *confidence limits*, where such tables are concerned, also arise in cases where the null hypothesis is discarded. To introduce these let us look again at the general case in which the members of a sample from a given population are classified according to the presence (1) or absence (0) of two attributes x and y. Suppose that for a sample of size n the observed side frequencies in the table are as shown in Table 1.14 and that X members of the sample are found to possess both attributes (the reason for the terminology used will be explained later).

TABLE 1.14

	Attribute x		
	(0)	(1)	Total
Attribute y { (0)			$(n-y)$
(1)		X	y
Total	s_2	s_1	n

Now, if the side frequencies in the table are taken as *fixed*, and X is considered to be a random approximately normally distributed variable, we may estimate the standard error of the 'true' mean of X in the population. Then at the risk of being wrong in the long run a small number of times (say once in twenty times—the 95 per cent confidence limits) we may use this standard error to set up lower and upper limits for the mean of X in repeated samples of size n having the same side frequencies.

The statistical problems involved in obtaining confidence limits in the case of fourfold tables have been considered by several writers, in particular by Cornfield (1956) and Cox (1958). Unfortunately the underlying theory is too complicated to be reproduced here even in a simplified form. But since answers to the problems concerned are of considerable practical importance it is desirable to show—even if it is only at the arithmetic level—how they may be obtained. The procedures outlined here follow Cox's presentation (for Cornfield's methods, though equally attractive from a theoretical viewpoint, involve the iterative solution of a quartic equation) and to facilitate the student who wishes to consult his paper his terminology is used. The answers arrived at are approximate only, but in cases where none of the expected frequencies in the fourfold table is small the approximations appear to be good.

An example. Suppose that of the members of a given population equally exposed to a virus infection a percentage—which we will assume contains a fair cross-section of the population as a whole—have been inoculated. After the epidemic has passed a random sample of people from the population is drawn and the numbers of inoculated and uninoculated in it that have escaped infection are recorded and the figures in Table 1.15 obtained.

TABLE 1.15

Incidence of Virus Infection

		Not infected	Infected	Total
Not inoculated	. .	54	48	102
Inoculated	. .	35	11	46
Total	. .	89	59	148

Comparing this table with Table 1.14 we see that n, the sample size, is 148; y is 46, X is 11 and s_1 is 59. To estimate the confidence limits of the 'true' mean of X we proceed as follows. First we calculate the proportion (m_1) in the sample that was infected by the virus; we also calculate the variance of this proportion. They are given respectively by the formulae:

$$m_1 = s_1/n \quad \text{and} \quad m_2 = m_1(1-m_1) \qquad \ldots 1.12$$

and for our data have the values $0 \cdot 397$ and $0 \cdot 239$. The estimate of the variance of the true mean of X is given by the expression:

$$\sigma_{o,y}^2 = \frac{y(n-y)m_2}{(n-1)} \qquad \ldots 1.13$$

Substituting the required values in this expression its numerical value is found to be $7 \cdot 628$, hence $\sigma_{o,y}$, the standard error of the true mean of X, is $\sqrt{(7 \cdot 628)} = 2 \cdot 762$. With an observed value of X of 11 the 95 per cent confidence limits of the true mean, assuming a normal distribution, are therefore:

$$11 \pm (1 \cdot 960 \times 2 \cdot 762)$$

namely $5 \cdot 586$ and $16 \cdot 414$. If the sample size is small a continuity correction can be applied simply by subtracting $\frac{1}{2}$ from the lower imit and by adding $\frac{1}{2}$ to the upper limit. The adjusted limits now are 5·086 and 16·914.

From these results we conclude that in repeated sampling from this population, the sample size being 148, the number of people that were infected though they had been inoculated would, 95 times out of 100 on the average, lie between 5 and 17. Another way of expressing the results, which is perhaps easier to appraise, is to express them in terms of relative risks.

Relative risks. It is clear from the data in Table 1.15 that the proportion of uninoculated people that was affected by the virus is considerably larger than the proportion of inoculated affected. In other words the risk of being affected had you been inoculated is less than the risk had you not been inoculated. It will now be shown how these relative risks can be expressed in quantitative terms.

The arithmetic procedure involves solving for β the two linear equations formed by letting the expression $m_{o,y} + \beta\sigma^2_{o,y}$ equal, in turn, the lower and the upper limits for the 'true' mean of X, stated above. In this expression the value of $\sigma^2_{o,y}$ has already been found; the value of $m_{o,y}$ is given by the product ym_1 and is 18·262 for the data in Table 1.15. On solving these equations the two values of β are found to be $-0·177$ and $-1·727$ for the lower and the upper limits respectively. Next the expression $\psi = e^\beta$ is evaluated for each of the β-values, e being the constant 2·718, and this leads to the required results, namely 0·178 and 0·838. We are now in a position to say that at the 95 per cent confidence limits, that is with a chance of being wrong on the average once in twenty times, the risk that an inoculated person will be infected by the virus is at most 83·8 per cent of that for an uninoculated person and it may be as low as 17·8 per cent.

Summary and Discussion

In this chapter the use of the chi-square test as a test of association in fourfold contingency tables is discussed. The Yates correction for increasing the validity of the test when small samples are involved is described. Next Fisher's exact test for finding the probability of any obtained configuration of cell frequencies in 2×2 tables, when the side frequencies are taken as fixed, is illustrated and the method is shown to provide a useful check on the accuracy of results obtained by the chi-square test.

As the amount of calculation involved when the exact test is used is onerous a word of direction to the student as to when to use each of the alternative tests will be helpful. Cochran (1954) recommends that Fisher's exact test should be used when the total sample size in a 2×2 table is less than 20, or when the sample is less than 40 and one of the expected frequencies is less than 5. In other cases the chi-square test corrected for continuity may be relied upon to give satisfactory results.

Next the problem of comparing frequencies in matched samples is considered, and then—using the data already presented for illustrative purposes—a word of warning is given about possible sources of bias in the collection of frequency data. The chapter ends with a section on confidence limits where fourfold tables are concerned.

Finally, though the matter has not been mentioned in the text, the student's attention may be drawn to various tables and charts which exist to enable him to make quick, though sometimes only rough, tests of significance where fourfold tables and pairs of proportions are concerned. In particular *Tables for use with Binomial Samples*, by Mainland *et al.* (1956), may be mentioned. In cases where tests of significance for a long series of 2×2 tables are required these tables will prove of great value as many of the answers can be obtained by direct reference to them.

REFERENCES

BERKSON, J. (1946) 'Limitations of the application of the fourfold table analysis to hospital data', *Biometrics*, **2**, 47–53

COCHRAN, W. G. (1954) 'Some methods of strengthening the common χ^2 tests', *Biometrics*, **10**, 417–51

COMRIE, L. J. (1952) *Barlow's Tables*, London, E. & F. N. Spon Ltd.

COX, D. R. (1958) 'The regression analysis of binary sequences', *Journal of Royal Statistical Society*, **B20**, 215–32

FISHER, R. A. (1950) *Statistical Methods for Research Workers*, Edinburgh, Oliver & Boyd

FISHER, R. A. & YATES, F. (1957) *Statistical Tables for Biological, Agricultural and Medical Research* (5th ed.), Edinburgh, Oliver & Boyd

HALD, A. (1952) *Statistical Tables and Formulas*, New York, Wiley & Sons

KENDALL, M. G. (1952) *The Advanced Theory of Statistics*, Vol. I, London, Griffin & Co.

LINDLEY, D. V. & MILLER, J. C. P. (1953) *Cambridge Elementary Statistical Tables*, Cambridge, University Press

MCNEMAR, Q. (1955) *Psychological Statistics*, New York, Wiley & Sons

MAXWELL, A. E. (1958) *Experimental Design in Psychology and the Medical Sciences*, London, Methuen & Co. Ltd.

STUART, A. (1957) 'The comparison of frequencies in matched samples', *British Journal of Statistical Psychology*, **10**, 29–32

YATES, F. (1934) 'Contingency tables involving small numbers and the χ^2 test', Supplement to the *Journal of the Royal Statistical Society*, **1**, 217–35

FURTHER READING

ARMITAGE, P. (1960) 'The construction of comparable groups,' in *Controlled Clinical Trials*, ed. A. B. Hill, Blackwell, Oxford

CORNFIELD, J. (1956) 'A statistical problem arising from retrospective studies', in *Proceedings of the Third Berkeley Symposium*, Vol. IV, edited by J. Neyman, Berkeley, University of California Press

TAYLOR, I. & KNOWELDEN, J. (1957) *Principles of Epidemiology*, London, J. & A. Churchill Ltd.

READY RECKONER TABLES

MAINLAND, D., *et al.* (1956) *Tables for use with Binomial Samples*, Department of Medical Statistics, New York University College of Medicine

CHAPTER II

Contingency Tables with more than One Degree of Freedom

Introduction. In the last chapter 2×2 tables only were discussed. Now a description will be given of the use of χ^2 for testing for association when multiple, rather than dichotomous, classification categories are employed. In such cases the contingency tables have r rows and c columns where r or c, or both, are greater than 2. As already mentioned the degrees of freedom for an $r \times c$ contingency table are $(r-1)(c-1)$, in other words $(r-1)(c-1)$ cells in the body of the table can be filled arbitrarily. With such tables it is still widely believed that all the expected frequencies should be greater than 5 for a chi-square test to be reliable. However, Cochran (1954) and others have pointed out that this rule is too stringent. To relax the rule Cochran suggests that 'if relatively few expectations are less than 5 (say, one cell out of five or more, or two cells out of ten or more) a minimum expectation of 1 is allowable in computing χ^2'. This suggestion results from an examination of the exact distribution of the expression $\sum \{(O-E)^2/E\}$ which differs somewhat from the approximate distribution of it from which the χ^2 table is derived (Appendix, Table A).

When even the relaxed restrictions are not fulfilled resort can be had to other methods. These include an extension of Fisher's exact test. The latter, however, is so laborious that a 2×3 example only is considered in this book. A full discussion of the method can be found in an article by Freeman and Halton (1951). More practical methods for dealing with contingency tables such as Table 2.2 below, where most or all of the expected frequencies are less than 5, are arrived at by considering the exact mean and variance (Bartlett, 1937; Haldane, 1939) of χ^2. This approach is especially suitable when the number of

38

degrees of freedom approaches 30 or so; however, as will be seen, an ordinary chi-square test—when the number of degrees of freedom is only moderately large, say 15 or thereabouts—appears to be remarkably reliable even when most of the expected frequencies are as low as 1 or 2 and the result is referred to the tabulated values (Appendix, Table A). But let us begin with a straightforward example.

A 3 × 3 Contingency Table

In the following example the problem is one of testing whether there is an association between *type* and *site* of cerebral tumours. The tumours were divided into three types:

A	.	.	.	benignant tumours
B	.	.	.	malignant tumours
C	.	.	.	other cerebral tumours

The sites concerned were:

I	.	.	.	frontal lobes
II	.	.	.	temporal lobes
III	.	.	.	other cerebral areas

The incidence of the different tumours in the several sites for a sample of neurosurgical patients was:

TABLE 2.1

Incidence of Cerebral Tumours

Site	Type A	B	C	Total
I	23 (21·02)	9 (9·97)	6 (7·01)	38
II	21 (15·49)	4 (7·35)	3 (5·16)	28
III	34 (41·49)	24 (19·68)	17 (13·83)	75
Total	78	37	26	141

The numbers in brackets are the expected frequencies on the assumption of independence between site and type of tumour. Each expected frequency is obtained by dividing the product of the row total and the column total for the cell in question by the total frequency or sample size. For instance the expected frequency 21·02 is obtained from $(38 \times 78)/141$. From Table 2.1 it will be seen that once the four expected frequencies 21·02, 9·97, 15·49 and 7·35 have been obtained the others can be deduced since the marginal totals are known. It follows too that the 3×3 table has four degrees of freedom. This number could also have been obtained by substituting 3 for r and c in the formula $(r-1)(c-1)$.

The details of the calculations can now be arranged in a table similar to Table 1.4. This leads to nine expressions of the form $(O-E)^2/E$ which when evaluated and added give the required value of χ^2. The nine values are 0·19, 1·96, 1·35, 0·09, 1·53, 0·95, 0·15, 0·90, 0·72, the sum of which is 7·84. This value of χ^2 with four degrees of freedom is not significant so that no association between site and type of tumour can be claimed on the evidence obtained from these data.

A formula for calculating χ^2 for $r \times c$ contingency tables. It is generally quicker arithmetically to obtain the value of χ^2 for an $r \times c$ contingency table by a special formula. Suppose the entries in such a table are:

		Columns			
Rows	1	2	....	c	*Total*
1	a_1	a_2	....	a_c	t_1
2	b_1	b_2	....	b_c	t_2
.	.	.		.	.
.	.	.		.	.
.	.	.		.	.
r	h_1	h_2	....	h_c	t_r
Total	n_1	n_2	....	n_c	N

then

$$\chi^2 = N\left[\frac{1}{t_1}\sum_1^c \frac{a_i^2}{n_i} + \frac{1}{t_2}\sum_1^c \frac{b_i^2}{n_i} + \ldots + \frac{1}{t_r}\sum_1^c \frac{h_i^2}{n_i} - 1\right] \quad \ldots 2.1$$

For the data in Table 2.1:

$$\frac{1}{t_1}\sum_1^c \frac{a_i^2}{n_i} = \frac{1}{38}\left\{\frac{23^2}{78}+\frac{9^2}{37}+\frac{6^2}{26}\right\} = 0\cdot2725$$

$$\frac{1}{t_2}\sum_1^c \frac{b_i^2}{n_i} = \frac{1}{28}\left\{\frac{21^2}{78}+\frac{4^2}{37}+\frac{3^2}{26}\right\} = 0\cdot2297$$

$$\frac{1}{t_3}\sum_1^c \frac{c_i^2}{n_i} = \frac{1}{75}\left\{\frac{34^2}{78}+\frac{24^2}{37}+\frac{17^2}{26}\right\} = 0\cdot5534$$

so that:

$$\chi^2 = 141(0\cdot2725+0\cdot2297+0\cdot5534-1)$$
$$= 7\cdot84, \text{ as obtained above.}$$

The Exact Mean and Variance of χ^2

These statistics will now be considered and their use, when testing for association in large contingency tables in which the expected frequencies are small, will be demonstrated. But first a few comments about the tabulated values of χ^2. In the earlier editions of Fisher and Yates's *Statistical Tables* values of χ^2 were tabulated up to thirty degrees of freedom only, though in the latest edition (1957) of these tables, and in other standard publications—for instance, *Biometrika Tables for Statisticians* by Pearson and Hartley—tabulated values at intervals up to one hundred degrees of freedom are now available. However, Fisher and Yates pointed out at the outset that for n degrees of freedom, where n is greater than about 30, the expression $\sqrt{(2\chi^2)}$ is approximately normally distributed with mean of $\sqrt{(2n-1)}$ and standard deviation of unity. It follows that to test the significance of a value of χ^2 based on more than about thirty degrees of freedom the expression:

$$\sqrt{(2\chi^2)}-\sqrt{(2n-1)}$$

can be calculated and treated as a normal deviate with mean of zero and standard deviation of unity.

It is well though to remember that the procedure just mentioned holds only on the assumption that the expected frequencies are fairly large. When the expected frequencies in a contingency table are universally small then, though χ^2 is still approximately normally distributed for n greater than 30 or so, its mean and variance are no longer the same as those underlying the tabulated values of χ^2; in short neither the tabulated values nor the expression

$$\sqrt{(2\chi^2)} - \sqrt{(2n-1)}$$

are altogether reliable for assessing the significance of an obtained χ^2. To overcome the difficulty Haldane (1939) has given expressions for the exact mean (given earlier by Bartlett, 1937) and variance of χ^2 under these special conditions, i.e. when the degrees of freedom are greater than about 30 and the expected frequencies are in general very small. Under these conditions the expression for the mean of χ^2, denoted by $E(\chi^2)$, is:

$$\frac{(r-1)(c-1)N}{N-1} \qquad \ldots 2.2$$

where r stands for rows and c for columns in the contingency table, and N is the sample size. The expression for the variance is very forbidding in appearance, but an equivalent and simpler expression has been given by Dawson (1954). It is:

$$V(\chi^2) = \frac{2N}{N-3}(n_1-u_1)(n_2-u_2)+\frac{N^2}{N-1}u_1u_2 \qquad \ldots 2.3$$

where

$$n_1 = (r-1)(N-r)/(N-1); \; n_2 = (c-1)(N-c)/(N-1);$$
$$u_1 = (N\sum R_i^{-1}-r^2)/(N-2); \text{ and } u_2 = (N\sum C_j^{-1}-c^2)/(N-2)$$

In the latter expressions R_i stands for the sum of the entries in the i-th row, and C_j for the sum of the entries in the j-th column of the $r \times c$ contingency table.

The use of these expressions for testing for association, using the data in Table 2.2, will now be illustrated. (Regarding these data it is worth mentioning that since the two variables, age and lie-scale ratings, are continuous the problem of testing for association could also be tackled (more efficiently) by methods to be outlined in Chapter 4, but when the classification categories do not represent continuous scales the present approach is needed.) In the table ratings on a five-point 'lie scale' for 95 girls in the age range 5 to 15 years are given. Rating 5 refers to inveterate liars while rating 1 refers to those girls who, as far as was known, did not tell lies.

TABLE 2.2

Girls' Ratings on a Lie Scale

Age in years	Ratings					Total
	5	4	3	2	1	
5	0	0	2	0	1	3
6	0	0	0	1	3	4
7	0	1	0	3	0	4
8	1	1	3	1	0	6
9	0	2	3	2	2	9
10	0	2	7	6	0	15
11	0	2	2	6	1	11
12	2	0	2	5	5	14
13	0	1	2	5	4	12
14	1	3	3	2	0	9
15	1	2	3	2	0	8
Total	5	14	27	33	16	95

The table has 11 rows and 5 columns so that a value of χ^2 calculated from the data would have $10 \times 4 = 40$ degrees of freedom. It is clear too from the small size of the sample and the small marginal frequencies that practically all the expected frequencies fall below the value 5. It is possible then that a chi-square test applied in the usual way would not give a very reliable test of association between age and the girls' ratings on the lie-scale, but the data are suitable for an application of expressions 2.2 and 2.3 above for the exact mean and the exact variance of χ^2, and using these statistics a test based on the normal distribution can be performed.

First the value of χ^2 is calculated for the data in the usual way. It is found to be $51 \cdot 042$. Using formula 2.2 the exact mean of χ^2 is found to be:

$$E(\chi^2) = \frac{10 \times 4 \times 95}{94} = 40 \cdot 426$$

Moreover, using the expressions given above we find:

$$n_1 = 8 \cdot 936 \qquad n_2 = 3 \cdot 830$$
$$u_1 = 0 \cdot 39418 \quad \text{and} \quad u_2 = 0 \cdot 14108$$

On substituting these values in equation 2.3 we find:

$$V(x^2) = 70 \cdot 40221$$

The square root of this value, namely $8 \cdot 390$, is the standard deviation of x^2.

All the information needed for testing for an association between age and the lie ratings is now available. The required critical ratio is:

$$
\begin{aligned}
\text{C.R.} &= \frac{x^2 - E(x^2)}{\sqrt{[V(x^2)]}} \\
&= (51 \cdot 042 - 40 \cdot 426)/8 \cdot 390 \\
&= 1 \cdot 265
\end{aligned}
\qquad \ldots 2.4
$$

This value is referred to the normal curve; it corresponds to a probability of $0 \cdot 206$ which is non-significant. When the original chi-square value of $51 \cdot 042$, with forty degrees of freedom, is substituted in the formula $\sqrt{(2x^2)} - \sqrt{(2n-1)}$ the normal deviate $1 \cdot 216$ is obtained. This value corresponds to a probability of $0 \cdot 224$. The latter agrees very well with the more accurate value of $0 \cdot 206$. This finding is encouraging for it suggests that the chi-square test applied in the ordinary way to contingency tables such as Table 2.2 might be expected to give remarkably reliable results—in the latter table only one of the 55 expected frequencies is greater than 5.

When testing for significance using the exact mean and variance of x^2 Professor Bartlett has suggested that a more accurate assessment than that obtained from equation 2.4 would probably be derived by the use of a modified x^2, say x'^2. It is obtained by the formula

$$x'^2 = 2Ax^2/B \qquad \ldots 2.5$$

with $n' = 2A^2/B$ degrees of freedom, where A stands for the expected mean and B for the expected variance of x^2.

Combining Classification Categories

When the number of degrees of freedom are nothing like as large as in the last example, and when Cochran's relaxed rule, which it will be recalled allowed one expected frequency in five, two in ten, and so on to fall below five, is not met, resort can sometimes be had to the technique of combining classification categories. No general rules for this procedure are possible for the nature of the data being analysed, as well as the statistical considerations underlying the validity of the χ^2 test, must be considered. An example in which data for a number of groups of psychiatric patients are analysed will illustrate the non-statistical aspects of the problem.

Example. In the table below the incidence of hallucinations amongst patients from six different diagnostic categories are given and the investigator wishes to test if there is an association between the symptom and the psychiatric classifications.

TABLE 2.3

Incidence of Hallucinations amongst Psychiatric Patients

Diagnostic classification	Symptom present	Symptom absent	Total
Schizophrenics	16 (7·25)	24 (32·75)	40
Obsessionals	0 (3·99)	22 (18·01)	22
Hysterics	9 (6·35)	26 (28·65)	35
Anxiety states	0 (3·63)	20 (16·37)	20
Affective disorders	7 (7·25)	33 (32·75)	40
Neurotic depressives	1 (4·53)	24 (20·47)	25
Total	33	149	182

The entries in the brackets in the table are the expected frequencies on the assumption of independence. Three of these expected frequencies fall below five, but none is very small. The value of χ^2 calculated in the usual way is 26·918 which with five degrees of freedom is very highly significant, P being less than 0·001. With so high a level of significance an association can confidently be claimed between the symptom and the psychiatric diagnoses under consideration. However, if the fact that three of the expected frequencies fall below 5 is disturbing it would be well to examine the problem and see if the data could be regrouped in some meaningful way which would avoid the difficulty. In this example regrouping is possible for Schizophrenics and Affective Disorders may be thought of as forming a wider classification of psychiatric patients, namely Psychotics, while the remaining groups may be classified as Neurotics.

Rearranging the data in this way Table 2.4 is obtained.

TABLE 2.4

Data in Table 2.3 Regrouped

			Symptom present	Symptom absent	Total
Psychotics	.	.	23	57	80
Neurotics	.	.	10	92	102
Total	.	.	33	149	182

From the table it is seen that while 23/80, or 28·8 per cent, of psychotics show the symptom only 10/102, or 9·8 per cent, of neurotics do so. To test whether these percentages differ χ^2 is calculated in the usual way. It is found to be 10·842 which with one degree of freedom is significant, the probability being less than 0·001. This result confirms that reached earlier and incidentally lends further support to Cochran's contention that 'the expected frequencies greater than 5' rule can be relaxed for contingency tables greater than 2×2.

Though it is convenient on occasions to regroup data or combine categories to enable a χ^2 test to be performed a considerable amount of information may be lost in this way. Indeed the procedure might detract greatly from the interest and usefulness of a study. In the last example for instance, though most psychiatrists would agree to the combination of schizophrenic patients and patients with affective disorders on the one hand, and the remaining groups on the other, an examination of the data in Table 2.3 suggests that where hallucinations are concerned Hysterics should not be thrown in with the Neurotics in general. For this reason, and bearing in mind that if a large proportion of the expected frequencies is small then the χ^2 test may be very misleading, resort should again be had to exact methods. To illustrate the latter an example will now be worked out in detail. However, as mentioned earlier, the calculations are very onerous and we will confine ourselves to the case of a 2×3 contingency table.

The Exact Test for 2×3 Contingency Tables

In this example the problem is to assess the probability of an association between type and site of tumour in the dominant hemisphere.

The sites are, (I) the frontal lobes, and (II) the temporal lobes, while the types of tumours or lesions are (A) benign tumours, (B) malignant tumours, and (C) other cerebral tumours. The data, for a sample of 37 neurosurgical cases, are given in the table below.

TABLE 2.5

Lesions in the Dominant Hemisphere

Site of lesion	Type of lesion			Total
	A	B	C	
I	11 (13·14)	6 (3·89)	1 (0·97)	18
II	16 (13·86)	2 (4·11)	1 (1·03)	19
Total	27	8	2	37

The numbers in brackets in the table are the expected frequencies, on the assumption of independence between the two classifications, for instance $4·11 = (8 \times 19)/37$. In this example four of the expected frequencies fall below 5, two of them being approximately unity. It is unlikely then that an ordinary chi-square test would yield a very reliable result. Of course it could be argued that little if any information would be lost if the C-category were omitted, or if categories A (benign tumours) and C (other tumours) were combined, in which case a 2×2 table would result. However, for illustrative purposes, category C will be retained.

Applying a chi-square test in the usual way a value of 2·908 is obtained, which with two degrees of freedom corresponds to a probability of 0·238 and would not be considered significant. As a consequence no association between type and site of tumour could be claimed. Now let us proceed to a more accurate analysis.

The marginal frequencies in Table 2.5 are:

TABLE 2.6

Marginal Frequencies in Table 2.5

	A	B	C	Total
I		b	c	18
II				19
Total	27	8	2	37

The first task is to write out all the possible arrays of cell frequencies which would give these marginal frequencies. Since the table has two degrees of freedom it follows that once we have allotted frequencies to two of the cells, say to b and c—though any two cells not in the same column could be chosen—the frequencies for the other cells follow, that is can be deduced from our knowledge about the marginal values. Considering cells b and c there are definite limits to the values which may be allotted to these letters. Clearly both b and c can be zero, but while b may range from zero to 8, c can range from zero only to 2. These facts follow from the knowledge that whereas the sum of the B-column is 8, that of the C-column is only 2.

All the possible pairs of values of b and c may now be written out. They are (putting b first in each pair):

and

(0 0), (1 0), (2 0), (3 0), ... (8 0);

and

(0 1), (1 1), (2 1), (3 1), ... (8 1);

and

(0 2), (1 2), (2 2), (3 2), ... (8 2);

twenty-seven pairs in all.

Using these pairs of values for b and c and the marginal frequencies in Table 2.6, twenty-seven arrays of cell frequencies can be written out. The first three are:

	A	B	C	A	B	C	A	B	C	Total
I	18	0	0	17	1	0	16	2	0	18
II	9	8	2	10	7	2	11	6	2	19
	27	8	2	27	8	2	27	8	2	37

Now denoting the cell frequencies and the marginal frequencies in the 2×3 contingency table by symbols, as below:

TABLE 2.7

	A	B	C	Total
I	a	b	c	R_1
II	d	e	f	R_2
	C_1	C_2	C_3	N

an expression for the probability of any specific array, corresponding to equation 1.7, can be written down. It is:

$$p = \frac{R_1!.R_2!.C_1!.C_2!.C_3!}{N!.a!.b!.c!.d!.e!.f!} \qquad \ldots 2.6$$

Or, if binomial coefficients are preferred, the expression for p becomes:

$$p = \frac{_{(a+d)}C_a \times _{(b+e)}C_b \times _{(c+f)}C_c}{_NC_{R_1}} \qquad \ldots 2.7$$

Substituting the values in the first array on page 48 in either of these expressions:

$$p_1 = 0 \cdot 000265$$

where the subscript refers to the number of the array. Substituting the values in the second array, on p. 48, we find:

$$p_2 = 0 \cdot 003819$$

In this manner all the twenty-seven p-values could be obtained. However, it is seen on inspection that:

$$p_2 = p_1 \frac{18 \times 8}{1 \times 10}$$

Similarly,

$$p_3 = p_2 \frac{17 \times 7}{2 \times 11}$$

and

$$p_4 = p_3 \frac{16 \times 6}{3 \times 12}, \text{ etc.}$$

Again on coming to the second set of arrays p_{10} is calculated using equations 2.6 or 2.7, while p_{11} can be obtained from p_{10} by the equation

$$p_{11} = p_{10} \frac{17 \times 8}{1 \times 11}$$

and so on. The check on the calculations is that the sum of the twenty-seven values of p must equal unity. In this example the twenty-seven p-values are as follows,

p_1 to p_9	p_{10} to p_{18}	p_{19} to p_{27}
0·000265*	0·000955*	0·000738*
0·003819*	0·011804*	0·007869*
0·020657*	0·055085	0·031780*
0·055085	0·127119	0·063560
0·079449	0·158899	0·068856
0·063559	0·110170	0·041314*
0·027542*	0·041327	0·013366*
0·005902*	0·007638*	0·002122*
0·000477*	0·000530*	0·000127*

The sum of the twenty-seven p-values is 1·000012, which is very close to the true value of unity. The p-value corresponding to the observed frequencies (Table 2.5) is the seventh in the second column, namely 0·041327. To obtain the overall probability value of a set of observed frequencies as extreme or more extreme than the set given in Table 2.5 Freeman and Halton advise us to take the sum of all the probability values found above which are equal to or less than the p-value for the observed frequency table. These values are marked with an asterisk and their sum is 0·218 when rounded to three decimal places.

We are now in a position to assess the accuracy of the χ^2 value of 2·908 found by applying the chi-square test in the usual way. This value corresponds to a probability of 0·238, and is directly comparable with the probability of 0·218 found by the exact test. This is so because 0·218 is the sum of the probability for the observed table and of all probabilities less than it; in other words it corresponds to a two-tail test. The result is very encouraging: despite the fact that, in Table 2.5, four of the expected frequencies are less than 5, two of them considerably so, the chi-square test still gives a very good approximation to the exact probability.

REFERENCES

BARTLETT, M. S. (1937) 'Properties of sufficiency and statistical tests', *Proceedings Royal Society*, **A160**, 268–82

COCHRAN, W. G. (1954) 'Some methods of strengthening the common χ^2 tests', *Biometrics*, **10**, 417–51

DAWSON, R. B. (1954) 'A simplified expression for the variance of the χ^2-function on a contingency table', *Biometrika*, **41**, 280

FISHER, R. A. & YATES, F. (1957) *Statistical Tables for Biological, Agricultural and Medical Research* (5th ed.), Edinburgh, Oliver & Boyd

FREEMAN, G. H. & HALTON, J. H. (1951) 'Note on an exact treatment of contingency, goodness of fit and other problems of significance', *Biometrika*, **38**, 141–9

HALDANE, J. B. S. (1939) 'The mean and variance of χ^2 when used as a test of homogeneity, when expectations are small', *Biometrika*, **31**, 346–55

PEARSON, E. S. & HARTLEY, H. O. (1956) *Biometrika Tables for Statisticians*, Vol. 1, Cambridge University Press

CHAPTER III

Partitioning the Degrees of Freedom in Contingency Tables

Introduction. In the case of a fourfold table, such as Table 1.1, the statistical interpretation of the outcome of a chi-square test is clear since the test may be viewed as one between just two proportions. The situation, however, is not so clear-cut with contingency tables having more than one degree of freedom. Here the comparison of more than two proportions is concerned and though a significant overall χ^2 would tell us that these proportions were heterogeneous a more detailed analysis would be required to decide just where the significant differences lay. Some methods which are useful when carrying out the more detailed analysis will be described in this and in the following chapter. They are concerned with the subdivision of the overall chi-square value into additive components or, to put it in another way, with the partitioning of the degrees of freedom on which the overall value is based. We will begin by discussing the $2 \times N$ contingency table.

Subdivision of degrees of freedom in $2 \times N$ tables. In Table 3.1 frequency data for the general $2 \times N$ contingency table are represented by sym-

TABLE 3.1

The General $2 \times N$ Contingency Table

Sample	A's	not-A's	Total	Proportion of A's
1	x_1	$n_1 - x_1$	n_1	$p_1 = x_1/n_1$
2	x_2	$n_2 - x_2$	n_2	$p_2 = x_2/n_2$
.	.	.	.	.
.	.	.	.	.
.	.	.	.	.
N	x_N	$n_N - x_N$	n_N	$p_N = x_N/n_N$
Total	T_x	$T - T_x$	T	$\hat{p} = T_x/T$

52

bols. In the last column of the table the estimated proportion of A's in each of a number of samples is shown, while in the final row of the table the estimated proportion, $\hat{p}$, for the combined samples appears. When data for a $2 \times N$ table have been tabulated in this way the overall chi-square value for the data can be obtained conveniently by the formula:

$$\chi^2 = \frac{\sum x_j p_j - \hat{p} T_x}{\hat{p} \hat{q}} \qquad \ldots 3.1$$

with $N-1$ degrees of freedom, where j runs from 1 to N and $\hat{q}$ is equal to $(1-\hat{p})$.

This method of computation, due to Brandt and Snedecor, is useful when there is some question of a change in level in the proportions as might be the case if the samples were arranged in a special order. Moreover, if it is desired to test whether the value of p is different for the first N_1 rows of the table from its value in the subsequent N_2 rows, where $N_1 + N_2 = N$, formula 3.1 can be applied separately to these subsets of rows and to the totals for each subset. In particular —as will be demonstrated in the following example—the overall value of chi square based on $N-1$ degrees of freedom can be partitioned into independent additive components as follows:

Components of χ^2 due to:	_Degrees of freedom_
(i) difference between p's in the two subsets	1
(ii) variation among p's in the first N_1 rows	$N_1 - 1$
(iii) variation among p's in the last N_2 rows	$N_2 - 1$
Total χ^2	$N-1$

Illustrative example. In a social investigation carried out retrospectively samples from five different urban areas, A to E, were concerned. Before an analysis of the available data for the samples was undertaken it was decided to check if the proportion of employed adults classified as 'partly skilled or unskilled workers' was the same in each sample. This check seemed to be especially important since two of the areas,

D and E, were in south-eastern England where the proportion of such workers is known to be smaller than in the country as a whole.

The data required for the check are tabulated in Table 3.2, the proportion of partly skilled and unskilled workers appearing in the final column.

TABLE 3.2

Partly Skilled and Unskilled Workers versus Others

Area sampled	Workers x_j	Others $n_j - x_j$	Total n_j	Proportion of workers p_j
A	80	184	264	0·303030
B	58	147	205	0·282927
C	114	276	390	0·292308
D	55	196	251	0·219124
E	83	229	312	0·266026
Total	390	1032	1422	

$$\hat{p} = 0\cdot274262$$
$$\hat{q} = 1 - \hat{p} = 0\cdot725738$$
$$\hat{p}\hat{q} = 0\cdot199042$$

Examination of the data shows that, as suspected, the proportions of partly skilled and unskilled workers in samples D and E are smaller than in the other samples. The statistical problem then is to see whether the differences are greater than might reasonably be attributed to sampling fluctuations. The analysis of the data proceeds as follows. First an overall chi-square value, based on four degrees of freedom is obtained; this will serve as a check on subsequent calculations. The overall value is then partitioned into three components as illustrated in the preceding section.

At the bottom of Table 3.2 the values of $\hat{p}, \hat{q}$, and $\hat{p}\hat{q}$ are given. These estimates will be used in the denominator of equation 3.1 throughout the calculations. For the overall chi square the numerator in equation 3.1 is given by:

$$(80)(0\cdot303030) + \ldots + (83)(0\cdot266026) - (390)(0\cdot274262) = 1\cdot145076$$

This result when divided by $0\cdot199042$, the estimated value of $\hat{p}\hat{q}$, gives $5\cdot752937$ which is the required overall chi-square value, based on four degrees of freedom.

Next a component due to the differences between samples A to C on the one hand and D and E on the other is obtained. The data are tabulated as in Table 3.3:

TABLE 3.3

Samples A, B and C versus Samples D and E

Samples	Workers	Others	Total	Proportion of workers
$A+B+C$	252	607	859	0·293364
$D+E$	138	425	563	0·245115
Total	390	1032	1422	0·274262

For this component the numerator in equation 3.1 is:

$$(252)(0·293364) + (138)(0·245115) - (390)(0·274262) = 3·791418$$

while the denominator is $0·199042$ as before. This leads to a chi-square value of $3·976136$ based on one degree of freedom.

From the data in Tables 3.2 and 3.3 components of the overall chi-square value due to differences within the proportions for samples A, B and C on the one hand and D and E on the other can be obtained. The first is given by:

$$(80)(0·303030) + \ldots + (114)(0·292308) - (252)(0·293364)$$
$$= 0·047550$$

divided by $0·199042$ which equals $0·238894$; it is based on two degrees of freedom for three samples have been compared. The component of chi square due to the difference between the proportions for D and E is

$$(55)(0·219124) + (83)(0·266026) - (138)(0·245115) = 0·306108$$

divided by $0·199042$ which gives $1·537907$, based on one degree of freedom.

The three separate components of the overall chi square can now be set out as follows. The check on the calculations is that they add to $5·752937$ which is the original chi-square value.

Component of χ^2 due to:	χ^2	Degrees of freedom	Significance level
(i) difference between $A+B+C$ and $D+E$	3·976136	1	$P<0·05$
(ii) differences within A, B and C	0·238894	2	N.S.
(iii) difference within D and E	1·537907	1	N.S.
Total or overall chi square	5·752937	4	N.S.

From the results we see that though the overall test does not reveal any heterogeneity between the samples the components of chi square support the contention that the proportion of partly skilled and unskilled workers in areas D and E is smaller than in areas A, B and C, though within these two subgroups no differences exist.

More General Methods for Partitioning χ^2

The method outlined in the last section is not confined to cases where a set of samples is divided into just two subsets, but more general methods still exist. In particular Irwin (1949) and Lancaster (1949, 1950) have shown that the overall chi square for a contingency table can always be partitioned into as many components as the table has degrees of freedom. Their methods will be discussed below. An interesting consequence of these procedures is that it frequently happens that an overall χ^2, which itself is not significant, yields one or more components which are significant. The sensitivity of the test is thereby increased. To illustrate the point let us examine the following results regarding the incidence of 'retarded activity' in three samples of patients.

TABLE 3.4

Retarded Activity amongst Psychiatric Patients

	Affective disorders	Schizo-phrenics	Neurotics	Total
Retarded activity . .	12	13	5	30
No retarded activity .	18	17	25	60
Total . . .	30	30	30	90

An overall test for association on these data gives a value of $5\cdot700$ for χ^2, which with two degrees of freedom just falls short of the 5 per cent level of significance. An examination of the data, however, suggests that though the incidence of the symptom for the first two groups is very alike it occurs more frequently amongst these groups than in the neurotic group. At first one is tempted to combine the Affective Disorders and the Schizophrenics and do a χ^2 test on the resulting 2×2 table. Indeed if this is done a value of χ^2 equal to $5\cdot625$ is obtained, which with one degree of freedom is significant just beyond the $2\cdot5$ per cent level. But such a procedure, carried out purely in the hope of achieving a significant result, after the overall χ^2 test had failed to yield one, would be quite unjustified and contrary to good statistical practice. At any rate the procedure whereby categories are combined is one resorted to only reluctantly when expected frequencies are so small as to cause concern about the validity of the χ^2 test. In Table 3.4 the expected frequencies are all 10 or more so that there is no justification for combining groups on this pretext. Of course had we decided to combine the first two groups before examining the data then everything would have been in order though we would, by this process, have lost one degree of freedom unnecessarily. Luckily methods of partitioning the overall χ^2 value provide us with means of examining our data in greater detail and of obtaining more sensitive tests of association than we could have obtained otherwise.

Reference has already been made to the work of Irwin and Lancaster on the partitioning of the degrees of freedom of χ^2, but the calculating procedures illustrated below are due to Kimball (1954) who has supplied convenient formulae for obtaining the required results. To introduce them let us consider the following contingency table having r rows and c columns.

TABLE 3.5

					Total
a_1	a_2	a_3	$\ldots$	a_c	A
b_1	b_2	b_3	$\ldots$	b_c	B
c_1	c_2	$\ldots$	$\ldots$		C
.	.	.		.	.
.	.	.		.	.
n_1	n_2	n_3.		n_c	N

The value of chi square computed in the ordinary way from this table has $(r-1)(c-1)$ degrees of freedom and the first step in sub-division is to construct the $(r-1)(c-1)$ fourfold tables from which the components of chi square are calculated. In the case of a 2×3 table, for example, the two fourfold tables may be constructed as follows:

a_1	a_2		a_1	a_2	a_3
b_1	b_2		b_1	b_2	b_3

where the entries in each quadrant are added. For instance, using the data in Table 3.4 the corresponding fourfold tables are

12	13		25	5
18	17		35	25

At this stage the reader may ask why we combine the first two columns in preference to any other combination. The answer to this question must be supplied by the investigator; he is free to combine those two columns which are likely to be most meaningful in the light of his prior knowledge about the classification categories concerned, but the decision about which columns are to be combined should always be made before examining the data which he is about to analyse.

For a 3×3 contingency table there are four degrees of freedom and one way of constructing the four fourfold tables is this:

a_1	a_2		a_1	a_2	a_3
b_1	b_2		b_1	b_2	b_3

a_1	a_2		a_1	a_2	a_3
b_1	b_2		b_1	b_2	b_3
c_1	c_2		c_1	c_2	c_3

The formulae given by Kimball for a 2×3 table for the two components χ_1^2 and χ_2^2 of the total χ_T^2, where $\chi_T^2 = \chi_1^2 + \chi_2^2$ and the symbols are as defined by Table 3.5, are:

$$\chi_1^2 = \frac{N^2(a_1 b_2 - a_2 b_1)^2}{ABn_1 n_2(n_1 + n_2)} \qquad \ldots 3.2$$

$$\chi_2^2 = \frac{N[b_3(a_1 + a_2) - a_3(b_1 + b_2)]^2}{ABn_3(n_1 + n_2)} \qquad \ldots 3.3$$

On substituting the values in Table 3.4 in these formulae we get

$$\chi_1^2 = \frac{90^2(12 \times 17 - 13 \times 18)^2}{30 \times 60 \times 30 \times 30 \times 60} = 0 \cdot 075$$

and

$$\chi_2^2 = \frac{90(25 \times 25 - 5 \times 35)^2}{30 \times 60 \times 30 \times 60} = 5 \cdot 625$$

These two component values of χ^2 add to the value $5 \cdot 700$ of the overall χ^2. Each component has one degree of freedom and whereas the first is not significant the second is significant beyond the $2 \cdot 5$ per cent level. Partitioning of the overall χ^2 value, which itself was not significant, has given us a more sensitive test and we are now in a position to say that whereas the first two groups of patients do not differ where the symptom 'retarded activity' is concerned, the two groups combined differ significantly from the third group.

Partitioning χ^2 in a $2 \times c$ contingency table. Kimball's formulae for partitioning the total χ_T^2 value in the case of a $2 \times c$ contingency table are obtained by giving t the values 1, 2, . . ., $(c-1)$ in turn in the general formula:

$$\chi_t^2 = \frac{N^2[b_{t+1} S_t^{(a)} - a_{t+1} S_t^{(b)}]^2}{ABn_{t+1} S_t^{(n)} S_{t+1}^{(n)}} \qquad \ldots 3.4$$

where

$$S_t^{(a)} = \sum_{i=1}^{t} a_i, \qquad S_t^{(b)} = \sum_{i=1}^{t} b_i, \qquad S_t^{(n)} = \sum_{i=1}^{t} n_i$$

It is well to note that formulae 3.2, and 3.4 differ slightly from the ordinary formula (Chapter 1) for χ^2 for a 2×2 contingency table, in as far as they have an additional term in the denominator while instead of N in the numerator they have N^2. These formulae, as they stand, contain no correction for continuity but in cases where such a correction is desirable it can be applied in the usual way. When this is done additivity is no longer exact but the discrepancy between the overall χ^2 and the sum of its component parts in general is negligible.

Partitioning χ^2 in the 3×3 contingency table. The data in Table 3.6 will be used to illustrate the calculations in the formulae given by Kimball for partitioning the total χ_T^2 in a 3×3 table. The data refer to a sample of 95 girls within the age range 5–15 years attending a clinic for maladjusted children. The girls were rated on a nine-point scale for sensitivity, where the extremes of the scale stood for 'over-sensitive' and 'callous' respectively. Later the data were regrouped as shown in the Table 3.6.

TABLE 3.6

Girls Rated on a Sensitivity Scale

Age range (years)	Over-sensitive	Normal	Callous	Total
5–9	12 (a_1)	9 (a_2)	5 (a_3)	26 (A)
10–12	12 (b_1)	25 (b_2)	3 (b_3)	40 (B)
13–15	14 (c_1)	11 (c_2)	4 (c_3)	29 (C)
Total	38 (n_1)	45 (n_2)	12 (n_3)	95 (N)

To facilitate the calculations the appropriate letter from Table 3.5 is entered beside each cell frequency, and along the margins. Kimball's formulae for the 3×3 case, which has four degrees of freedom, are:

$$\chi_1^2 = \frac{N[B(n_2 a_1 - n_1 a_2) - A(n_2 b_1 - n_1 b_2)]^2}{ABn_1 n_2 (A+B)(n_1 + n_2)}$$

$$\chi_2^2 = \frac{N^2[b_3(a_1 + a_2) - a_3(b_1 + b_2)]^2}{ABn_3(A+B)(n_1 + n_2)}$$

$$\chi_3^2 = \frac{N^2[c_2(a_1 + b_1) - c_1(a_2 + b_2)]^2}{Cn_1 n_2 (A+B)(n_1 + n_2)}$$

$$\chi_4^2 = \frac{N[c_3(a_1 + a_2 + b_1 + b_2) - (a_3 + b_3)(c_1 + c_2)]^2}{Cn_3(A+B)(n_1 + n_2)}$$

The overall χ^2 value for the data in Table 3.6 is $6 \cdot 875$, which is well below $9 \cdot 49$ the value required with four degrees of freedom for significance at the 5 per cent level. Substituting the data from the table in the four formulae just given we get:

$$\chi_1^2 = \frac{95[40(45 \times 12 - 9 \times 38) - 26(45 \times 12 - 38 \times 25)]^2}{26 \times 40 \times 38 \times 45 \times 66 \times 83} = 3 \cdot 366$$

$$\chi_2^2 = \frac{95^2(3 \times 21 - 5 \times 37)^2}{26 \times 40 \times 12 \times 66 \times 83} = 1 \cdot 965$$

$$\chi_3^2 = \frac{95^2(11 \times 24 - 14 \times 34)^2}{29 \times 38 \times 45 \times 66 \times 83} = 1 \cdot 493$$

$$\chi_4^2 = \frac{95(4 \times 58 - 8 \times 25)^2}{29 \times 12 \times 66 \times 83} = 0 \cdot 051$$

$$\text{Total} = 6 \cdot 875$$

The values of the four components into which the overall χ^2 has been partitioned add up exactly to the value of the overall χ^2 itself, which yields a check on the calculations. Of the components the first is significant almost at the 5 per cent level. This component corresponds to the fourfold table:

a_1	a_2		12	9
b_1	b_2	i.e.,	12	25

and suggests that the proportion of our sample of girls in the age range 10–12 years who are normal on the sensitivity scale is greater than the proportion in the age range 5–9 years.

Kimball also gives a general formula for finding the components of the overall χ^2 in the case of the $(r \times c)$-fold table but as it is very cumbersome the reader is referred to the original article should he require it.

REFERENCES

IRWIN, J. O. (1949) 'A note on the subdivision of χ^2 in certain discrete distributions', *Biometrika*, **36**, 130–4

KIMBALL, A. W. (1954) 'Short-cut formulas for the exact partition of χ^2 in contingency tables', *Biometrics*, **10**, 452–8

LANCASTER, H. O. (1949) 'The derivation and partition of χ^2 in certain discrete distributions', *Biometrika*, **36**, 117–29

LANCASTER, H. O. (1950) 'The exact partition of χ^2 and its application to the problem of pooling of small expectations', *Biometrika*, **37**, 267–70

CHAPTER IV

Testing for Trends in Contingency Tables

Introduction. In the last chapter attention was drawn to the difficulties of interpretation which arise when a chi-square test, performed on a contingency table having more than one degree of freedom, is significant. Methods of overcoming these difficulties, by subdividing the overall chi-square value, were then described and it was noted that this subdivision often increased the sensitivity of the test. In this chapter a further method of subdividing the overall chi-square value will be considered. It is applicable to contingency tables where the classification categories fall into a natural order for then it is possible to search for trends in the data, in particular a component of chi square due to a linear trend can be separated out and tested for significance. A limiting feature of the method is that it should be confined to contingency tables where the categories into which the data are classified cover a relatively narrow 'range'. The five age categories in Table 4.1, which cover the age range 5 to 15 years only, afford an example. However, this feature is not very restrictive.

Classifications which fall into a natural order. In the examples discussed so far—with the exception of the data in Table 2.2—no natural order is detectable within the categories into which the data are classified. In Table 1.1 for instance it is immaterial as regards the chi-square test which sex is placed first. Again in Table 2.1 the result obtained is not affected if the order of the letters *A*, *B* and *C*, which refer to types of tumour, are rearranged, or if the order in which the sites of tumour, I, II and III, are arranged is altered. But in other contingency tables— Table 2.2 for example—the categories have a natural order. When this is the case the data can in general be treated as if they referred to

quantitative variables, as indeed they often do, and the theory of
linear regression can be applied to them.

To illustrate the use of regression methods for partitioning the
overall chi-square value data for a sample of 223 boys, attending the
same clinic as the girls referred to in Table 2.2, will be considered.
Like the girls the boys too were rated on a five-point scale as regards
lying. Before giving the data, however, one comment is necessary.
When planning the investigation to which the data in Tables 2.2 and
4.1 refer, it appeared reasonable to try to rate the children on a five-
point scale. However, the psychiatrist who did the rating came to the
conclusion that such a scale was inappropriate for it appeared that
the children belonged to one of two categories, *a* inveterate liars
(corresponding to grades 4 and 5 on the five-point scale), and *b* those
who in general refrained from lying (grades 1, 2 and 3 on the scale).
It was decided therefore when analysing the boys' data to use a
two-point scale only. The data appear in the following table:

TABLE 4.1

Boys' Ratings on a Lie Scale

		Age groups in years					
	Score	(5–7) −2	(8–9) −1	(10–11) 0	(12–13) 1	(14–15) 2	Total
Inveterate liars	1	6	18	19	27	25	95
Non-liars.	0	15	31	31	32	19	128
Total		21	49	50	59	44	223

An overall χ^2 test on the frequencies in this table gave a value of
$6 \cdot 691$ which with four degrees of freedom is not significant, hence we
might conclude that there was no association between age and lying.
However, if the proportion of inveterate liars in each age group is
found, namely:

$$0 \cdot 286 \quad 0 \cdot 367 \quad 0 \cdot 380 \quad 0 \cdot 458 \quad 0 \cdot 568$$

it is clear immediately that the proportions increase steadily with age.
The problem now is one of partitioning the overall χ^2 value so that
this trend in the proportions can be examined statistically. Moreover
since a trend, or regression line, is based on just one degree of freedom

it is possible that, though the overall χ^2 is not significant, the trend may be.

The problem of partitioning χ^2 when the data are ordered has been discussed by Yates (1948), Armitage (1955) and Cochran (1954). The procedure briefly is as follows. When there exists a natural order amongst the categories in a classification it may be assumed that there is a continuous variable underlying them. Under this assumption it is possible to quantify the variables by allotting numerical values to the categories. This has been done in Table 4.1 where the age groups (5–7) to (14–15) have been allotted scores running from −2 to +2. The lie scale too has been quantified by allotting the value +1 to the category 'inveterate liars' and the value 0 to 'non-liars'. These quantitative values are chosen quite arbitrarily. They are evenly spaced in the present example but they need not be. For instance if it were thought that lying was especially associated with puberty and the immediate post-pubertal period then the scores for age might be taken as −2, −1, 0, 3, 6, or some other values which gave greater weight to the year-groups in which we were especially interested. The reason for choosing −2 as the first score has no other significance than that it helps to keep the arithmetic simple since the five scores −2 to +2 add to zero.

Having quantified our data we may now proceed to treat them as we would data for a bivariate frequency table and calculate the correlation between the two variates 'age' and 'lying', or the regression of one of these variates on the other. For this purpose it will be convenient to express the formulae for the regression coefficients in the form which involves the *difference* between the variates (*vide*, Aitken, 1947, p. 92) for this will clarify the calculations. The formula for estimating the regression coefficient of y' (lying) on x' (age) is:

$$b_{yx} = \{\textstyle\sum x^2 + \sum y^2 - \sum (x-y)^2\}/2\sum x^2 \qquad \ldots 4.1$$

and that for x' on y' is:

$$b_{xy} = \{\textstyle\sum x^2 + \sum y^2 - \sum (x-y)^2\}/2\sum y^2 \qquad \ldots 4.2$$

while the variances of these two regression coefficients, on the null hypothesis, are respectively:

$$V(b_{yx}) = s_y^2/\textstyle\sum x^2 \quad \text{and} \quad V(b_{xy}) = s_x^2/\textstyle\sum y^2 \qquad \ldots 4.3$$

In these equations the x and y values refer to deviations from the means of the respective variates, while s_y^2 is an estimate of the variance of the y-variate and s_x^2 one of the variance of the x-variate.

To evaluate the expressions 4.1, 4.2, and 4.3 for the data in Table 4.1 it is convenient to draw up four frequency distributions from which the sums of squares about their means of the variates x', y', $(x'-y')$ and $(x'+y')$ may be obtained. The sum of squares of the latter variate allows a check to be made on the calculations.

Taking the variates x' and y' first the frequency distributions and the data required to obtain the sum of squares about the means of these two variates are as follows:

TABLE 4.2

Frequency Distribution for y'					Frequency Distribution for x'			
y'	f	fy'	fy'^2		x'	f	fx'	fx'^2
1	95	95	95		2	44	88	176
0	128	0	0		1	59	59	59
					0	50	0	00
					−1	49	−49	49
					−2	21	−42	84
Total	223	95	95		Total	223	56	368

The sum of squares of the y'-values about their mean now is:

$$\sum y^2 = 95 - 95^2/223 = 54\cdot529$$

while that for the x'-values is:

$$\sum x^2 = 368 - 56^2/223 = 353\cdot937$$

The frequency distributions for $(x'-y')$ and $(x'+y')$ are obtained as follows. The frequency 6 in the top left-hand corner of the table has an $(x'-y')$ score of $(-2-1)$, that is of -3: the frequencies 18 and 15 each have an $(x'-y')$ score of -2, and so on. Indeed if the values of $(x'-y')$ and $(x'+y')$ are calculated for the complete table we obtain:

$$(x'-y')$$
$$\begin{array}{ccccc} -3 & -2 & -1 & 0 & 1 \\ -2 & -1 & 0 & 1 & 2 \end{array}$$

$$(x'+y')$$
$$\begin{array}{ccccc} -1 & 0 & 1 & 2 & 3 \\ -2 & -1 & 0 & 1 & 2 \end{array}$$

hence the frequency distributions are those shown in Table 4.3.

TABLE 4.3

Frequency Distribution for $(x'\ y')$				*Frequency Distribution for $(x'+y')$*			
$(x'-y')$	f	$f(x'-y')$	$f(x'-y')^2$	$(x'+y')$	f	$f(x'+y')$	$f(x'+y')^2$
3	6	−18	54	3	25	75	225
−2	33	−66	132	2	46	92	184
−1	50	−50	50	1	51	51	51
0	58	0	0	0	49	0	0
1	57	57	57	−1	37	−37	37
2	19	38	76	−2	15	−30	60
Total	223	−39	369	Total	223	151	557

From these results we obtain:

$$\sum (x-y)^2 = 369 - (-39)^2/223 = 362 \cdot 179$$

and
$$\sum (x+y)^2 = 557 - 151^2/223 = 454 \cdot 754$$

The check on the calculations is that:

$$2\left(\sum x^2 + \sum y^2\right) = \sum (x-y)^2 + \sum (x+y)^2$$

that is:

$$2(54 \cdot 529 + 353 \cdot 937) = 362 \cdot 179 + 454 \cdot 754$$

The estimates of the variances of x' and y' can now be obtained. They are respectively:

$$s_x^2 = 353 \cdot 937/223 = 1 \cdot 58716$$

and
$$s_y^2 = 54 \cdot 529/223 = 0 \cdot 24453$$

All the information required for obtaining the estimates of the two regression coefficients and their variances is now available. On substituting it in equations 4.1 to 4.3 we find:

$$b_{yx} = (353 \cdot 937 + 54 \cdot 529 - 362 \cdot 179)/(2 \times 353 \cdot 937) = 0 \cdot 065389$$
$$b_{xy} = (353 \cdot 937 + 54 \cdot 529 - 362 \cdot 179)/(2 \times 54 \cdot 529) = 0 \cdot 424425$$
$$V(b_{yx}) = 0 \cdot 24453/353 \cdot 937 = 0 \cdot 00069089$$

and

$$V(b_{xy}) = 1 \cdot 58716/54 \cdot 529 = 0 \cdot 029107$$

The two regression coefficients may now be tested for significance. This is done in one of two equivalent ways; (1) by finding the critical ratio $b/\sqrt{V(b)}$ which gives:

$$0\cdot065389/\sqrt{0\cdot00069089} = 2\cdot4877 \quad \text{for } b_{yx}$$
and
$$0\cdot424425/\sqrt{0\cdot029107} = 2\cdot4877 \quad \text{for } b_{xy}$$

These ratios are referred to the normal curve. Or (2) by finding the squares of the critical ratios, which, since the regression coefficients are based on one degree of freedom will be distributed as χ^2. The latter method is more appropriate for present purposes since the χ^2 value is being partitioned. The required χ^2 values are:

for b_{yx}: $\chi^2 = (0\cdot065389)^2/0\cdot00069089 = 6\cdot188713$
and for b_{xy}: $\chi^2 = (0\cdot424425)^2/0\cdot029107 = 6\cdot188713$

An interesting fact, pointed out by Yates, is now seen for whether the regression coefficient b_{yx} or b_{xy} is used the χ^2 value obtained is the same.

Recalling that the overall χ^2 for the data in Table 4.1 is $6\cdot691$ based on four degrees of freedom the following table can now be drawn up:

Source of variation	Degrees of freedom	χ^2	Probable level
Due to linear regression	1	$6\cdot189$	$P < 0\cdot025$
Departure from regression line (by subtraction)	3	$0\cdot502$	N.S.
Overall value	4	$6\cdot691$	N.S.

It is seen then that though the overall χ^2, with four degrees of freedom, is not significant, the χ^2 value due to regression, based on only one degree of freedom, is significant beyond the $2\cdot5$ per cent level. Partitioning the overall value has therefore greatly increased the sensitivity of the test and, returning to the data, Table 4.1, we conclude that there is a significant increase in lying with increase in age for the age range in question. We can further say that the increase is linear rather than curvilinear for departure from linear regression is represented by a chi square value of only $0 \cdot 502$, based on three degrees of freedom, which is a long way from being significant.

Testing for Trend in an $r \times c$ Contingency Table

In a $2 \times c$ contingency table it is sufficient for the classification with the c categories to have a natural order if we wish to test for trend, for the other classification of the data, being a dichotomy, can then be quantified simply by allotting the numerical value 1 to one member of the dichotomy and the value 0 to the other. In the general contingency table with r rows and c columns, on the other hand, it is necessary for both classifications to have a natural order for trends to be tested by the regression method. One further example will help in getting the matter clear, though the calculations need not be worked out in such great detail as in the last example.

The sample of 223 boys already referred to were rated on a four-point scale for the symptom 'disturbed dreams', where a rating of 4 meant that the boy concerned suffered severely as a consequence of his dreams while a boy rated 1 did not suffer at all in this way. The values of the ratings themselves, that is the numbers 4, 3, 2, and 1, could have been retained as the quantified scale in the table below but to simplify the calculations as much as possible they were replaced by the numbers 2, 1, 0 and -1, where 2 refers to the boys with the most disturbed dreams. It is important for the student to note that such a change in the values allotted to the variable, provided the intervals between adjacent categories—in our case unity—do not change, in no way affects the results obtained; in other words, the value of $\sum x^2$ will be the same whether we use the values 4, 3, 2 and 1, or the values 2, 1, 0 and -1, or any other set of four numbers differing one from the next by unity.

The data for the 223 boys with numerical values allotted to the categories are given in Table 4.4.

The overall χ^2 value for these data is $31 \cdot 670$, which, with twelve degrees of freedom, is significant beyond the 1 per cent level. We are interested, however, in testing whether there is a trend in the proportions as age increases and if so whether this trend is linear or of a more complicated nature. Considering 'ratings' as the x'-variate and 'age' as the y'-variate we can draw up frequency distributions for each, and for the variates $(x'-y')$ and $(x'+y')$; but as the latter is required

TABLE 4.4

Boys' Ratings for Disturbed Dreams

Age groups in years	Score	Ratings				Total
		2	1	0	−1	
5–7	−2	7	3	4	7	21
8–9	−1	13	11	15	10	49
10–11	0	7	11	9	23	50
12–13	1	10	12	9	28	59
14–15	2	3	4	5	32	44
Total		40	41	42	100	223

for checking purposes only it will be omitted here. The other three distributions and auxiliary data are given below:

(1)

x'	f	fx'	fx^2
2	40	80	160
1	41	41	41
0	42	0	0
−1	100	−100	100
Sum	223	21	301

(2)

y'	f	fy'	fy^2
−2	21	−42	84
−1	49	−49	49
0	50	0	0
1	59	59	59
2	44	88	176
Sum	223	56	368

From these two tables the values

$$\sum x^2 = 301 - 21^2/223 = 299 \cdot 022$$

and

$$\sum y^2 = 368 - 56^2/223 = 353 \cdot 937$$

are obtained.

To find the frequency distribution for $(x'-y')$ we can enter mentally the value $(x'-y')$ beside each of the frequencies in Table 4.4. For example the entry beside 7 in the top left-hand cell is $(2-(-2))$ or 4; those beside 3 and 13 are respectively $(1-(-2))$ and $(2-(-1))$ both of which are 3, hence we add 3 and 13 when constructing the frequency distribution for $(x'-y')$. In a similar manner the third entry in the frequency distribution is the sum of diagonal values 7, 11 and 4, that is 22, and so on. The distribution is:

(3)

$(x'-y')$	f	$f(x'-y')$	$f(x'-y')^2$
4	7	28	112
3	16	48	144
2	22	44	88
1	43	43	43
0	34	0	0
-1	36	-36	36
-2	33	-66	132
-3	32	-96	288
	223	-35	843

Hence $\sum (x-y)^2 = 843 - (-35)^2/223 = 837 \cdot 507$

From these results the required estimates are found to be:

$$b_{yx} = 0 \cdot 30858 \qquad b_{xy} = -0 \cdot 2607$$
$$V(b_{yx}) = 0 \cdot 00530784 \qquad V(b_{xy}) = 0 \cdot 0037885$$

Using either pair of results the value of χ^2, due to regression and based on one degree of freedom, is found to be $17 \cdot 9398$. The overall χ^2 value is now partitioned as follows:

Source of variation	Degrees of freedom	χ^2	Significant level
Due to linear regression	1	$17 \cdot 9398$	$P < 0 \cdot 001$
Due to departure from linear regression (by subtraction)	11	$13 \cdot 7307$	N.S.
Overall value	12	$31 \cdot 6705$	$P < 0 \cdot 005$

We have now demonstrated that the highly significant overall chi-square value is due in large part to a linear trend in the proportions in the cells of the contingency table. To get a clearer picture of the significance of this finding let us turn to the regression coefficients themselves. Both are negative, but that fact in itself has no particular significance until we consider the direction in which the two variates have been scored. Taking age first we note that a high score corresponds to older children, a low score to younger children. Taking 'disturbed dreams' we note that a high score corresponds to a very disturbed state, a low score to an undisturbed state. The fact that the regression coefficients are negative now tells us that the condition of having disturbed dreams is one associated with younger boys; or, put in another way, as boys grow older they tend to be annoyed less and less by disturbed dreams.

REFERENCES

AITKEN, A. C. (1947) *Statistical Mathematics*, Edinburgh, Oliver & Boyd

ARMITAGE, P. (1955) 'Tests for linear trends in proportions and frequencies', *Biometrics*, **11**, 375–86

COCHRAN, W. G. (1954) 'Some methods of strengthening the common χ^2 tests', *Biometrics*, **10**, 417–51

YATES, F. (1948) 'The analysis of contingency tables with groupings based on quantitative characters', *Biometrika*, **35**, 176–81

CHAPTER V

Combining and Comparing Results from Different Investigations

Introduction. When a number of independent, but similar, investigations have been carried out, tests of significance performed and probability values obtained, the probability of the combined results is sometimes required. One way of obtaining it is to evaluate the expression:

$$\chi^2 = -4 \cdot 605 \sum \log_{10} P_i \qquad \ldots 5.1$$

where there are n separate probabilities P to be combined. The result is then referred to the χ^2 table with $2n$ degrees of freedom. But this method is rather crude for it gives each investigation equal weight though the sample sizes may be very different. Furthermore the test does not take into account the directions in which the discrepancies from expectation in the separate investigations occur. For these reasons it is necessary to consider what other methods are available for combining results and when they can be used to the best advantage. In keeping, however, with the warning against bias given in Chapter 1 it is to be understood that the combination of data from different investigations, or of parameters estimated from such data, must be undertaken only in cases where the investigations are true replications of each other, or when differential effects from one investigation to another can be ruled out. The discussion will begin by considering an example in which it is legitimate to combine the raw data themselves for several contingency tables.

Combining two or more contingency tables. In Table 3.1 the incidence of the symptom 'retarded activity' in a random sample of 90 psychiatric patients, 30 from each of the diagnostic categories Affective

Disorders, Schizophrenics and Neurotics is given. These data reappear as Sample I in Table 5.1, together with similar data for another sample of 120 patients, 40 from each of the diagnostic categories concerned. At the bottom of the table the data for the two samples are combined.

In the table the proportion of patients showing the symptom in each diagnostic group for each sample is given. Since corresponding proportions are very alike the samples may safely be combined and an overall analysis of the data performed. The value of χ^2 for the

TABLE 5.1

Retarded Activity amongst Psychiatric Patients

Sample		Affective disorders	Schizo-phrenics	Neurotics	Total
I	Retarded activity	12 (0·400)	13 (0·433)	5 (0·167)	30
	No retarded activity	18	17	25	60
		30	30	30	90
II	Retarded activity .	17 (0·425)	15 (0·375)	5 (0·125)	37
	No retarded activity	23	25	35	83
		40	40	40	120
I + II	Retarded activity .	29	28	10	67
	No retarded activity	41	42	60	143
	Total . . .	70	70	70	210

2×3 table of combined frequencies is $15 \cdot 036$. With two degrees of freedom this value corresponds to a probability $P < 0 \cdot 001$, showing that there is a strong association between the presence of the symptom and the diagnostic categories in question. Examination of the data leaves little doubt that the presence of 'retarded activity' in Neurotics is much less frequent than in Affective Disorders and Schizophrenics, while the incidence of the symptom in these two categories is roughly equal. To verify this conclusion the overall χ^2 value of $15 \cdot 036$, which we will call χ_T^2, can be partitioned—as shown in Chapter 3—into two independent and additive components χ_1^2 and χ_2^2, the former providing a comparison between the two psychotic categories, the latter providing a comparison between the 140 psychotic patients on the one hand and the 70 neurotics on the other.

The results are shown below and, for completeness, the overall χ^2 values for the two samples themselves have been partitioned and are given for purposes of comparison.

	χ_1^2	χ_2^2	χ_T^2
Sample I	0·075 (N.S.)	5·625 (2%)	5·700 (N.S.)
Sample II	0·234 (N.S.)	9·456 (1%)	9·690 (1%)
Samples I & II	0·033 (N.S.)	15·003 (0·1%)	15·036 (0·1%)

The results strongly support the conclusion already reached that the contrast in the data is between Neurotics on the one hand and Affective Disorders and Schizophrenics on the other; in each instance χ_1^2 is not significant while χ_2^2 is highly so. Combining the data for the two samples too increases the significance levels.

Another, but less efficient, method of procedure with the above results would have been to add the separate χ^2 value for the two samples and refer the result to the tables with two degrees of freedom. The sum of the two values of χ_1^2 is 0·309 and is not significant. The sum of the two values of χ_2^2 is 15·081 but, though with two degrees of freedom this value is still highly significant, it is not as pronounced as the χ_2^2 value of 15·003 with one degree of freedom got from combining the contingency tables.

Taking the Direction of Discrepancies into Account in 2×2 Tables

The square root of χ^2, which naturally is denoted by χ, when based on just one degree of freedom is approximately normally distributed when the expected frequencies are greater than about 5. It can assume positive and negative values, has a mean of zero and a standard deviation of unity. Furthermore the algebraic sum of n independent values of χ is approximately normally distributed with mean of zero and standard deviation equal to $\sqrt{n}$. This fact can be utilised when combining results from a number of fourfold tables and it has the advantage that it allows the signs of discrepancies from expectation to be taken into account. The method is especially appropriate when the sample sizes in the separate investigations are roughly equal (say in the ratio 2 to 1) and the proportions being compared lie approximately within the range 0·2 to 0·8.

To illustrate the method let us consider the data in Table 5.2 in which the incidence of malignant and benignant tumours in the left and right hemispheres in the cortex are given. The problem is to test whether there is an association between hemisphere and type of tumour. Data for three sites in each hemisphere were available, but an earlier investigation had shown that there was no reason to suspect that the relationship if any between hemisphere and type of tumour would differ from one site to another so that an overall assessment of the hemisphere-tumour relationship was indicated. For each of the three sites the number of patients—32, 27 and 34 respectively—are

TABLE 5.2

Incidence of Tumours in the Two Hemispheres for Different Sites in the Cortex

Site of tumour		Benignant tumours	Malignant tumours	Proportion of malignant tumours	χ^2	χ
I	Left hemisphere	15	6	0·286	0·002248	−0·04741
	Right hemisphere	8	3	0·272		
		23	9			
II	Left hemisphere	12	3	0·200	0·096429	0·31053
	Right hemisphere	9	3	0·250		
		21	6			
III	Left hemisphere	10	6	0·375	0·006919	0·08318
	Right hemisphere	11	7	0·389		
		21	13			

roughly equal. However, on examining the data it is seen that while the proportion of patients with malignant tumours in the left hemisphere for Site I slightly exceeds that in the right hemisphere, the reverse is true for Sites II and III. A test which takes this difference in the signs of the discrepancies into account is therefore indicated.

To do the test the value of χ^2, omitting the continuity correction, is calculated for each 2×2 table. The square roots of these values, indicated by χ, are then obtained. They appear in the final column of the table. A minus (or plus) sign is allotted to the value of χ for the first site where the proportion of malignant tumours in the left hemisphere exceeds that in the right, while a plus (or minus) sign is allotted to the χ-values for the other two sites where the discrepancies between the proportions are in the reverse direction. The test criterion for the combined results is given by the expression:

$$C.R. = \frac{-0\cdot04741 + 0\cdot31053 + 0\cdot08318}{\sqrt{3}}$$

$$= 0\cdot199936$$

This value is referred to the normal curve, but clearly it is not significant so that taking all three sites together there is no suggestion of an association between type of tumour and hemisphere.

Combining Results from 2×2 Tables having Very Unequal Sample Sizes—Cochran's Criterion

When the sample sizes, N_i, for a series of 2×2 tables are very different the procedure outlined in the last section is not efficient and a method of weighting the results is desirable. A very efficient method of doing this has been given by Cochran (1954). It would appear to be satisfactory for a wide range of values of N and even in cases where the proportions being compared in each contingency table are well outside the range $0\cdot2$ to $0\cdot8$. To illustrate Cochran's procedure we will consider the data in the following table where the incidence of tics in three age groups of boys and girls is given.

TABLE 5.3

The Incidence of Tics in Three Samples of Maladjusted Children

Age range			Tics	No tics	Total	Proportion with tics
5–9	Boys	.	13	57	70	0·1857
	Girls	.	3	23	26	0·1154
	Total	.	16	80	96	0·1667
10–12	Boys	.	26	56	82	0·3171
	Girls	.	11	29	40	0·2750
	Total	.	37	85	122	0·3033
13–15	Boys	.	15	56	71	0·2113
	Girls	.	2	27	29	0·0690
	Total	.	17	83	100	0·1700

In the last column of the table the proportion of boys and girls having tics in each age group is shown as is the proportion of the total number of children with tics in each group. The problem is to perform an overall test to see if the proportion of boys suffering from the complaint differs from the proportion of girls, and to do so with the knowledge that the number of boys and girls differs within each age group and from group to group and that the proportions vary considerably and one at least is very small.

The data required to do the test are tabulated below:

(1) d_i	(2) p_i	(3) $q_i = (1 - p_i)$	(4) $p_i q_i$	(5) w_i
0·0703	0·1667	0·8333	0·1389	18·96
0·0421	0·3033	0·6967	0·2113	26·89
0·1423	0·1700	0·8300	0·1411	20·59

The difference (d_i) between the pairs of proportions in each age group are given in column (1). In column (2) the overall proportion (p_i) in each group is given, and in column (3) the complements of these proportions, $q_i = (1 - p_i)$, appear. In column (4) the products of the corresponding p's and q's are given: these are the variances of the proportions p. Since the number of boys greatly exceeds the number of girls in each age group and the number in the age groups differ a weight to be applied to each sample is calculated. It is given by:

$$w_i = (n_{i_1} \times n_{i_2})/(n_{i_1} + n_{i_2}) \qquad \ldots 5.2$$

where n_{i_1} and n_{i_2} are the number of boys and girls respectively in the i-th sample. These weights appear in column (5). A weighted mean difference $\bar{d}$ is now obtained from the expression:

$$\bar{d} = \sum (w_i d_i)/w \qquad \ldots 5.3$$

where $w = \sum w_i$. This has a standard error:

$$\text{S.E.} = \sqrt{[(\sum w_i p_i q_i)]}/w \qquad \ldots 5.4$$

The test criterion required now is:

$$\text{C.R.} = \bar{d}/\text{S.E.} \qquad \ldots 5.5$$

For our data:

$$\sum w_i d_i = 0\cdot0703 \times 18\cdot96 + 0\cdot0421 \times 26\cdot89 + 0\cdot1423 \times 20\cdot59$$
$$= 5\cdot3949$$

and

$$\sum w_i p_i q_i = 0\cdot1389 \times 18\cdot96 + 0\cdot2113 \times 26\cdot89 + 0\cdot1411 \times 20\cdot59$$
$$= 11\cdot207$$

The critical ratio is therefore:

$$\text{C.R.} = 5\cdot3949/\sqrt{(11\cdot2209)} = 1\cdot61$$

Referring this value to the normal curve it is found to correspond to a probability of $0\cdot1074$. Had the three age groups been combined and an overall chi-square test performed a value of $2\cdot110$ would have been obtained. This corresponds to a probability of $0\cdot2838$ which is much larger than that given by Cochran's criterion. This fact illustrates the greater sensitivity of Cochran's test.

Comparing Contingency Tables

In investigations involving qualitative data it is often necessary to compare contingency tables and to summarise the information they provide. At first sight (see Table 5.4) such comparisons might appear to call for intricate statistical techniques, yet it often happens that the questions requiring to be answered can be dealt with adequately by simple methods. To demonstrate that this is so a rather complicated example will now be considered.

A social psychologist was interested in the effects of 'exposure to Western culture' on the attitudes of people in the Middle East. His samples—one of Moslems, the other of Christians—consisted of high-school students. Each sample was divided into three subgroups, a 'high', a 'medium' and a 'low' exposure group. The students in each subgroup had been asked to rate themselves on a three-point scale (1 standing for a good rating) regarding their own evaluation of their achievement at school, while their parents also rated them on a similar scale with respect to the same variable. The problem which the psychologist asked to have answered was whether the amount of agreement (if any) between parents' and students' ratings was the same in the three exposure groups, and also whether there were any clear-cut differences between Moslems and Christians. The data appear in Table 5.4.

Inspection of the six 3 × 3 contingency tables shows that there is a considerable measure of agreement between parents' and students' ratings, for the entries in each table tend to concentrate along the main diagonal. It is also clear from inspection that the other entries in the tables tend to lie above and to the right of the main diagonal rather than below and to the left; in other words, parents tend to rate their children higher than the children rate themselves. These are

TABLE 5.4

Ratings for Groups Exposed to Western Culture
(*students' ratings horizontal, parents' ratings vertical*)

		Moslems				Christians		
Group	Rating	1	2	3	Rating	1	2	3
High	1	27	48	5	1	42	40	1
exposure	2	4	91	12	2	4	96	5
	3	0	6	8	3	1	11	5
Medium	1	53	103	13	1	48	35	4
exposure	2	2	95	26	2	4	67	9
	3	1	9	12	3	1	8	8
Low	1	80	61	23	1	18	20	3
exposure	2	2	56	12	2	1	15	4
	3	3	10	17	3	1	0	5

general features of the data, but inspection does not reveal any noticeable differences either between the different exposure groups or between Moslems and Christians. To examine the data more carefully resort to statistical tests is indicated. However, it is not immediately clear at first just what comparisons should be made.

One approach would be to carry out a chi-square test for association separately on each contingency table and compare the results. This approach, however, would lead to difficulties for having obtained the separate χ^2 values there is no very satisfactory way of comparing them or of interpreting the results (Fairfield Smith, 1957). Another procedure would be to extract from the separate contingency tables particular rows, columns or particular entries which we wished to compare and perform separate tests of association on them. The results from this procedure might, in their turn, prove difficult to summarise succinctly. Perhaps the ideal would be achieved if one or two overall comparisons, which adequately answered the questions posed, could be found. On occasion such overall comparisons, as Fairfield Smith has demonstrated, do suggest themselves provided we get clear in our minds the questions we wish to answer.

In the present example the main questions posed by the psychologist are (a) whether the relationship between the parents' and the students' ratings differ from one exposure group to another, and (b) whether they differ from one religious group to another. To answer these questions we could, for example, count the number of agreements and disagreements between parents' and students' ratings at each exposure level, and for each religious group, and do chi-square tests for association on the tables which result. These counts are shown in Table 5.5, and the tests performed are discussed below. An alternative procedure, though in the case of the present data it might not give us much additional information, would be to count—in addition to the number of agreements—the number of children rated higher by their parents than by themselves, and the number rated lower. Other procedures too could easily be thought of and evaluated in the light of the data in hand and the questions to be answered.

TABLE 5.5

Agreement and Disagreement between Parents' and Childrens' Ratings (the proportions appear in brackets)

	Moslems		Christians	
	Agreements	Disagreements	Agreements	Disagreements
High exposure	126 (0·625)	75 (0·374)	143 (0·698)	62 (0·302)
Medium exposure	160 (0·509)	154 (0·491)	123 (0·668)	61 (0·332)
Low exposure	153 (0·579)	111 (0·421)	38 (0·567)	29 (0·433)
Total	439 (0·564)	340 (0·436)	304 (0·667)	152 (0·333)

In Table 5.5 the entry 126 is the sum of the diagonal entries in the first 3 × 3 table; the entry 75 is the sum of all the remaining entries in the same table. The other entries in Table 5.5 are obtained from the other 3 × 3 tables in a corresponding way.

Before comparing the religious groups it is advisable to carry out separate chi-square tests on each of the 2 × 3 contingency tables given in Table 5.5, otherwise interactions (Mood, 1946) between the groups, where the different 'exposures' are concerned, may be overlooked. In the case of Moslems the value of chi square is 7·276 which, with two degrees of freedom, is just significant at the 5 per cent level. Inspection of the proportions shows that agreement between parents and children is highest (62·6 per cent) in the 'high exposure' group, lowest in the 'medium exposure' group, and intermediate (57·9 per cent) in the 'low exposure' group.

For the 2 × 3 contingency table for Christians a value of chi square of 3·844 is obtained. It does not reach an acceptable level of significance so that different degrees of exposure to Western culture appear to have little or no effect on the amount of agreement between parents' and childrens' ratings in this case. It is worth noting, however, that in the case of Christians the proportion of agreements between the ratings increases steadily from 0·567 for the low exposure group to 0·698 for the high exposure group. A method for testing such a linear trend for significance was given in the last chapter.

Since it is now clear that the data for the two religious groups differ somewhat from one exposure group to another it would probably be best to compare the groups at each exposure level rather than to make a comparison between the total figures for the groups given at the bottom of Table 5.5. It would be interesting too to see whether more Moslems than Christians, or vice versa, are rated alike by parents and children under one rating than under another. To do the latter test the data in Table 5.6 can be extracted.

TABLE 5.6

Number Rated 1, 2 and 3 for Each Religious Group

		Ratings			
		1	2	3	Total
Moslems	.	160	242	37	439
Christians	.	108	178	18	304
Total	.	268	420	55	743

In this table the entry 160 is the number of students rated *1* both by themselves and by their parents in the Moslem group: 242 is the number rated *2*, and so on. When a chi-square test is performed on these data the value 1·94 is obtained. This value is based on two degrees of freedom and gives no reason to doubt that the two religious groups are alike as regards the numbers falling in the rating categories considered.

While the investigations performed in this section do not exhaust the number of interesting questions which could be asked with reference to the data in Table 5.4 they will, it is hoped, suffice to show how the required information may be extracted from a set of complicated contingency tables.

Comparing Trends in Contingency Tables

When, for two corresponding contingency tables, regression lines are fitted with a view to testing for trends in the data these lines themselves can be compared. In Table 4.4 data for 223 boys in five age groups rated for 'disturbed dreams' were given. The overall chi square, χ_T^2, for the resulting 5×4 table was then partitioned into two parts, one due to a linear trend in the proportions and found by calculating the regression of 'age' on 'disturbed dreams'.

In the table below similar data for 95 girls in the same age range and attending the same clinic as the boys are given.

TABLE 5.7

The Incidence of Disturbed Dreams in Girls

		Ratings				
Age groups	*Score*	*2*	*1*	*0*	*−1*	*Total*
5–7	−2	3	3	3	2	11
8–9	−1	3	0	4	8	15
10–11	0	2	2	8	14	26
12–13	1	6	3	6	11	26
14–15	2	4	2	2	9	17
Total		18	10	23	44	95

The overall chi square for this table was also calculated and later partitioned in a manner similar to that used on the boys' data. The results of the analysis for both sexes are tabulated below to facilitate comparison.

	N	b_{yx}	$V(b_{yx})$	χ_1^2	χ_{11}^2	χ_{12}^2
Boys	223	-0.309	0.00531	17.940 (0.1%)	13.731 (N.S.)	31.670 (0.5%)
Girls	95	-0.051	0.01232	0.215 (N.S.)	11.365 (N.S.)	11.580 (N.S.)

Here b_{yx} is the regression coefficient of age on 'disturbed dreams' (though, as was seen earlier, the regression coefficient b_{xy} might equally well have been used). $V(b_{yx})$ is the variance of the regression coefficient; χ_1^2 is the amount of the total χ^2 accounted for by linear regression, while χ_{11}^2 is the residual amount due to departure from linear regression. The subscripts indicate the number of degrees of freedom on which the chi square concerned is based. The significance levels are given in brackets.

For boys there is a significant association between 'age' and 'disturbed dreams' and a very significant linear trend in the proportions in the contingency table for them. For the girls no significant results are found. If we wish to test whether the regression coefficients for the boys' and the girls' data differ the test criterion is:

$$\text{C.R.} = (b_1 - b_2)/\sqrt{[V(b_1) + V(b_2)]}$$
$$= (0.309 - 0.051)/\sqrt{(0.00531 + 0.01231)}$$
$$= 1.94$$

This value, when referred to the normal curve, just falls short of the 5 per cent level of significance.

REFERENCES

COCHRAN, W. G. (1954) 'Some methods of strengthening the common χ^2 tests', *Biometrics*, **10**, 417–51

FAIRFIELD SMITH, H. (1957) 'On comparing contingency tables', *The Philippine Statistician*, **6**, 71–81

MOOD, A. M. (1946) Reply to query, *Biometrics*, **2**, 17–18

FURTHER READING

YATES, F. (1955) 'The use of transformations and maximum likelihood in the analysis of quantal experiments involving two treatments', *Biometrika*, **42**, 382–403

CHAPTER VI

The Analysis of 2^n Contingency Tables

Introduction. So far we have been concerned with two-dimensional contingency tables, that is tables in which the population sampled is classified—whether dichotomously, or using multiple categories—according to two characteristics. A logical extension of such classifications is to consider three or more characteristics simultaneously. For example the problem of testing for association in a 2^3 contingency table (that is one in which the data are dichotomised with respect to three characteristics) has been dealt with by Bartlett (1935). The same author indicates how this problem may be tackled by exact methods when the sample size, or one of the expected frequencies, is small. Freeman and Halton (1951) too, in an article referred to in Chapter 2, discuss in some detail the exact method for dealing with 2^3 contingency tables and give a worked example. However, the calculations involved either in doing a test of association or in calculating a probability by exact methods in the case of 2^3 tables are intricate, and in situations in which the null hypothesis is discarded the meaning of the results is often difficult to appraise. For $2 \times 2 \times 3$ contingency tables and tables of higher degree than the third problems of calculation and interpretation are more difficult still. Consequently, when dealing with classification problems involving more than two characteristics at a time it is generally more rewarding to seek a transformation of the data which will put the proportions being compared on a continuous linear scale and then to treat the transformed data by conventional methods for continuous variables. Examples of how this may be done in the case of 2^n tables are given by Winsor (1948) and by Dyke and Patterson (1952) and the procedure can be extended to include tables with multiple classification categories. The method described in the following example is that outlined by the latter writers.

Example. A psychiatrist is interested in the extent to which patients' recovery is predictable from the symptoms they show when ill. The presence (+) and absence (−) of three symptoms, *d* depressed, *a* anxious, and *g* delusions of guilt, in a sample of 819 male patients (in the age range 16–59 years) is noted. The tabulated data are as follows:

TABLE 6.1

Incidence of Symptoms amongst Psychiatric Patients

No. of patients	819							
Depressed (*d*)	345 (−)				474 (+)			
Anxious (*a*)	211 (−)		134 (+)		172 (−)		302 (+)	
Delusions (*g*)	205 (−)	6 (+)	128 (−)	6 (+)	139 (−)	33 (+)	216 (−)	86 (+)
Recovered	68/205	3/6	58/128	3/6	70/139	23/33	129/216	59/86
Proportion recovered	0·3317	0·5000	0·4531	0·5000	0·5036	0·6970	0·5972	0·6860

The procedure, were we to follow the methods of earlier chapters, would be to test for an association between the presence or absence of each symptom and the patients' recovery. But it would be more helpful still if, in addition to testing for such associations, an estimate of the effect of each variable—and of possible interactions between them—could be obtained.

Traditionally the analysis of data such as those in Table 6.1 has been approached by psychometricians along correlational lines. For instance the tetrachoric correlations between the three symptoms *d*, *a* and *g*, and between each and recovery, which we will call *c*, are:

Tetrachoric Correlation Matrix

	c	*d*	*a*	*g*
c		0·318	0·206	0·278
d			0·385	0·600
a				0·289

Using these correlations resort could be had to multiple regression methods and weights found to indicate the relative importance of the symptoms for predicting recovery. But in a correlation model of this kind the assumption is implicitly made that the variables involved are distributed in a multivariate normal distribution. However, such an assumption is rarely true where dichotomous variables are concerned, and tetrachoric correlations at all times are very unreliable (Kendall, 1949). Moreover, as Dyke and Patterson point out, correlational methods do not enable estimates of the effects of the separate variables (in our case, of the symptoms) to be obtained, while non-linear or interaction effects go undetected. In view of these objections it is desirable to approach the problem along other lines.

Transforming the data. Data arranged in a 2^n table, as ours are, immediately calls to mind factorial designs in analysis of variance (Maxwell, 1959, Chapter IV). The three symptoms could be looked on as three 'treatments', to use conventional language, each of which is in each instant either present or absent. With three treatments, each at just two levels, there are eight (2^3) combinations of treatments. The first is the case where each treatment is at the lower of its two levels; this corresponds in our case to the situation in which all the symptoms are absent; traditionally it is denoted by the symbol (*1*). A look at our data reveals that 205 of the 819 patients show none of the symptoms. Of these 205 patients 68, or a proportion equal to 0·3317, recover.

Reading along row four of Table 6.1 we find 6 patients in the next category. These are patients who are neither depressed nor anxious but who do suffer from delusions. Traditionally this class, or treatment combination, would be denoted by the symbol *g*, for we have denoted 'delusions' by the letter *g*. The next entry in row four of the table contains 128 patients, who though not depressed, were anxious and had delusions. This category is symbolically denoted by the letters *ag*; this simply means that while the symptoms *a* and *g* are present, *d* is not. Following this scheme of denotation—one familiar in experimental design (see Yates, 1937)—the eight combinational treatments represented by the entries in row four of Table 6.1 can be represented symbolically by:

$$(1) \quad g \quad a \quad ag \quad d \quad dg \quad da \quad dag$$

In row six of the table the proportion of patients who recovered in each category, that is under each of the eight combinations of treatments, is given. If these proportions are each near $0 \cdot 5$ and if the numbers on which each proportion is based are equal or roughly so then the data may be analysed by the familiar methods of analysis of variance as applied to factorial designs. When such a procedure is followed a linear additive model is assumed. For our data such an assumption would imply that the proportion of patients who recovered in each category was composed of independent additive components representing the effects of the several symptoms, and—if interactions between the symptoms were present—further components representing interaction effects. However, when the proportions being analysed differ widely from each other and from $0 \cdot 5$, and when they are based on very uneven numbers a straightforward analysis of them is likely to be misleading. This follows primarily from the fact that proportions since they lie between 0 and 1 are not continuous variables, nor is the assumption that they can be partitioned into additive components justified. Dyke and Patterson illustrate the latter point by an example. Suppose, they say, that two men firing at a target get bull's-eyes 10 and 5 per cent of the time respectively. Suppose too that under a change of conditions, such as lengthening the shooting range, the rate for the first man is reduced to 5 per cent, we do not expect a decrease of 5 per cent, that is to zero, in the rate for the second man. It would, however, be reasonable to expect a proportionate decrease, that is a decrease to $2\frac{1}{2}$ per cent in the second man's score, and any appreciable departure from proportionality would be described as an interaction between the men and the range. The problem then is one of finding a transformation of the proportions which would put them on a scale where the effects are likely to follow a linear law.

For proportions (p) a useful transformation is the logit transformation:

$$z' = \log_e p/(1-p) \qquad \ldots 6.1$$

If in this expression we substitute $r = 2p - 1$, and multiply each side of the equation by $\frac{1}{2}$ we get Fisher's z-transformation:

$$z = \tfrac{1}{2}\log_e\{(1+r)/(1-r)\} \qquad \ldots 6.2$$

which is tabulated—Table VII, Fisher and Yates, 1957. With this transformation the property of proportionality mentioned in the case of the two marksmen holds approximately even for small p.

We shall now apply this transformation to the data in Table 6.1, but as the numbers on which the proportions in that table are based are so unequal we shall not proceed to a straightforward analysis of the transformed data but rather, following Dyke and Patterson, use a maximum likelihood procedure to obtain best estimates of the effects sought, together with estimates of their standard errors.

Finding the transformed scores. The data are now tabulated as shown in Table 6.2. The symbols already described for denoting the presence or absence of the symptoms appear in column (1). In column (2) the number of patients in each category, that is under each treatment combination, appear. The proportion of patients in each category that recovered is given in column (3). These p values are then transformed into r values by the formula $r = 2p - 1$, and are given in column (4). The r-values are then transformed to z-values using Table VII, Fisher and Yates; these appear in column (5).

TABLE 6.2

Transforming the Proportions in Table 6.1

(1)	(2) n	(3) p	(4) $r = 2p - 1$	(5) z
(1)	205	0·3317	−0·3366	−0·3503
g	6	0·5000	0·0000	0·0000
a	128	0·4531	−0·0938	−0·0940
ag	6	0·5000	0·0000	0·0000
d	139	0·5036	0·0072	0·0072
dg	33	0·6970	0·3940	0·4165
da	216	0·5972	0·1944	0·1969
dag	86	0·6860	0·3720	0·3907

Since the z-values are approximately normally distributed we can proceed to analyse the data in column (5) by ordinary analysis of variance methods. This, for present purposes, is most conveniently done by the methods outlined by Yates (1937, p. 15). Details of the calculations appear in Table 6.3:

TABLE 6.3

Estimating Main Effects from the z-Scores in Table 6.2

(1)	(2)	(3)	(4)	(5)	(6)	Estimate
(I)	−0·3503	−0·3503	−0·4443	0·5670	0·070875	= m
g	0·0000	−0·0940	1·0113	1·0474	0·130925	= g
a	−0·0940	0·4237	0·4443	0·4202	0·052525	= a
ag	0·0000	0·5876	0·6031	−0·4718	−0·058975	
d	0·0072	0·3503	0·2563	1·4556	0·181950	= d
dg	0·4165	0·0940	0·1639	0·1588	0·019850	
da	0·1969	0·4093	−0·2563	−0·0924	−0·011550	
dag	0·3907	0·1938	−0·2155	0·0408	0·005100	
Total	0·5670					

The analysis yields estimates, *m*, *g*, *a* and *d*, of the true mean of the z-values and of the effects of the three symptoms; these appear in column (6): from the same column estimates of the interaction effects of the symptoms can also be read off if these are required, but in this analysis we shall content ourselves with finding estimates of the main effects only. Had the numbers of subjects in the categories been equal the above analysis would probably be adequate for all practical purposes, but as the numbers are unequal the precisions of the proportions differ, so that these first estimates of the main effects may not be very accurate. By the use of maximum likelihood methods Dyke and Patterson show how these estimates can be improved. This is done by an iterative process, but before going on, a word of explanation is required about the calculations in Table 6.3.

The entries in column (2) are the z-values from Table 6.2. Their sum is 0·5670, so that the mean of the eight readings is 0·070875—the first entry in column (6). The entries in column (2) are summed, two at a time, from the top downwards. The results give the first four entries in column (3). The entries in column (2) are also subtracted from each other, the first from the second, the third from the fourth, and so on; these differences form the last four entries in column (3). The entries in column (3) are now treated in the same way as those in column (2) were treated, and the results give the entries in column (4). The process is continued until a column is reached, the first entry of which is the sum of the original readings, in our case this occurs in column (5). As there is no efficient check on the calculations they should be repeated on a separate sheet and the results compared; this safeguard is essential for the method is very prone to error. The final column is then divided by the number of original readings—in our case 8—to obtain the estimates required. The latter, as already noted, appear in column (6).

Obtaining second approximations of the main effects. The entries in column (6) of Table 6.3 are first approximations of the main effects of the three symptoms under consideration. To obtain second approximations we proceed as follows. On the assumption that interaction effects are absent a column of values, which we will denote by Z', is obtained by the addition and subtraction of the first estimates of the mean and the main effects according to the symbolic notation for the treatment combinations used in Table 6.2. The first symbol is (1), which denotes that all the symptoms are absent, so that the first entry in the Z' column Table 6.4 is:

$$m - g - a - d = 0.070875 - 0.130925 - 0.052525 - 0.181950$$
$$= -0.294525$$

Table 6.4

Adjusted Treatment Values

(1)	(2) Z'	(3) R'	(4) $w' = (1 - R'^2)$	(5) z'	(6) nw'	(7) $nw'z'$
(1)	-0.2945	-0.2864	0.9180	-0.3492	188.1849	-65.7142
g	-0.0327	-0.0327	0.9989	-0.0327	5.9936	-0.1960
a	-0.1895	-0.1874	0.9649	-0.0925	123.5048	-11.4242
ag	0.0724	0.0723	0.9948	0.0003	5.9686	0.0018
d	0.0694	0.0693	0.9952	0.0070	138.3325	0.9683
dg	0.3312	0.3196	0.8979	0.4141	29.6292	12.2695
da	0.1744	0.1727	0.9702	0.1968	209.5578	41.2410
dag	0.4363	0.4104	0.8316	0.3901	71.5152	27.8981

The fourth symbol is ag, indicating that whereas the symptoms a and g are present, d is absent, so that the Z'-value for this row is:

$$m + g + a - d = 0.070875 + 0.130925 + 0.052525 - 0.181950$$
$$= 0.072375, \quad \text{and so on.}$$

The Z'-values are in turn referred to Table VII of Fisher and Yates's tables and equivalent r-values, which will we denote by R', obtained. From the R'-values a new z' column is obtained using the formula:

$$z' = Z' + (r - R')/(1 - R'^2) \qquad \ldots 6.3$$

These new values are known as the 'working z's' by analogy with probit analysis (see Mather, 1949, for a simple account). The expression by which they are obtained is derived by means of maximum likelihood considerations. In it $r = 2p - 1$, and it has maximal and minimal values of $Z' + 1/(1 + R')$ and $Z' - 1/(1 - R')$ for $p = 1$ and $p = 0$ respectively. Its use enables difficulties in the estimation procedure, which arise when some of the p-values are 1 or 0, to be bridged.

Weighted values of the z's are now found, as shown in columns (6) and (7), Table 6.4; w' is equal to $(1 - R'^2)$. The entries in the latter columns are now subjected to a process of addition and subtraction in pairs similar to that described for Table 6.3. The initial and final stages in this process are shown in Table 6.5.

TABLE 6.5

Estimating Weights and Weighted Effects

(1)	(2) nw'		(5)	(2') nw'z'		(5')
(*I*)	188·18	...	772·68	−65·71	...	5·04
g	5·99	...	−546·46	− 0·20	...	74·90
a	123·50	...	48·42	−11·42	...	110·39
ag	5·97	...	35·32	0·00	...	−78·74
d	138·33	...	125·40	0·97	...	159·71
dg	29·63	...	52·98	12·27	...	−78·98
da	209·56	...	177·82	41·24	...	1·42
dag	71·52	...	−94·00	27·90	...	29·45
Total	772·68			Total 5·04		

From the entries in columns (5) and (5') a matrix equation (6.4) is set up, which on solution will give more accurate estimates of the main effects: these are indicated by the symbols m', d', a' and g'. The 4×4 matrix on the left of equation (6.4) is formed from the entries in

column (5) of Table 6.5, omitting the entry $-94 \cdot 00$. The diagonal entries of the matrix are all equal to $772 \cdot 68$, while the remaining entries in the first row (and the first column of the matrix, since the latter is symmetrical) correspond to the symbols d, a and g in that order. The other cells of the matrix are filled by the first-order interaction effects in a way which is obvious on inspection. The entries in the column vector on the right of the equation come from column (5′) of Table 6.5 and correspond to the symbols (1), d, a and g.

$$
\begin{array}{cccc}
(1) & d & a & g
\end{array}
$$

$$
\begin{bmatrix}
772 \cdot 68 & 125 \cdot 40 & 48 \cdot 42 & -546 \cdot 46 \\
125 \cdot 40 & 772 \cdot 68 & 177 \cdot 82 & 52 \cdot 98 \\
48 \cdot 42 & 177 \cdot 82 & 772 \cdot 68 & 35 \cdot 32 \\
-546 \cdot 46 & 52 \cdot 98 & 35 \cdot 32 & 772 \cdot 68
\end{bmatrix}
\begin{bmatrix}
m' \\
d' \\
a' \\
g'
\end{bmatrix}
=
\begin{bmatrix}
5 \cdot 04 \\
159 \cdot 71 \\
110 \cdot 39 \\
74 \cdot 90
\end{bmatrix}
$$

$$\ldots 6.4$$

Solving this matrix equation is perhaps the most laborious part of the calculations. First the inverse of the 4×4 matrix is required. A number of methods for inverting a matrix are known (Dwyer, 1945; Thomson, 1951). The method described by the latter writer, due to Aitken, was used here and the inverse matrix was found to be:

$$
\begin{bmatrix}
0 \cdot 0028489 & -0 \cdot 0005729 & -0 \cdot 0001428 & 0 \cdot 0020666 \\
-0 \cdot 0005718 & 0 \cdot 0014863 & -0 \cdot 0002836 & -0 \cdot 0004941 \\
-0 \cdot 0001411 & -0 \cdot 0002835 & 0 \cdot 0013754 & -0 \cdot 0001436 \\
0 \cdot 0020604 & -0 \cdot 0004941 & -0 \cdot 0001437 & 0 \cdot 0027960
\end{bmatrix}
$$

This inverse matrix was then post-multiplied by the row matrix on the right of the matrix equation to give the required new estimates, namely, m', d', a' and g'. The value of m', for example, is:

$$m' = 0 \cdot 0028489 \times 5 \cdot 04 - 0 \cdot 0005729 \times 159 \cdot 71 - 0 \cdot 0001428$$
$$\times 110 \cdot 39 + 0 \cdot 0020666 \times 74 \cdot 90 = 0 \cdot 06188$$

The values of d', a' and g' are found in a similar manner.

The standard errors of these estimates are given by the square roots

of the diagonal entries of the inverse matrix above. The estimates of the main effects and of their standard errors are:

	Main effect	Standard error	C.R.	Significant level
$m' =$	0·06188	0·05338	1·159	N.S.
$d' =$	0·16619	0·03855	4·311	$P < 0·001$
$a' =$	0·09508	0·03708	2·564	$P < 0·02$
$g' =$	0·12503	0·05288	2·364	$P < 0·02$

Comparing the estimates with their standard errors gives a test of their significance. The results show that the d effect is highly significant, while the a and g effects, though significant, are less so. This means that prognosis is good for patients, one of whose symptoms is 'depression'. It is also fairly good for anxious patients and, to a slightly less extent, for patients suffering from delusions.

Iteration. If more accurate estimates still of the main effects are required the calculations could be repeated by forming a new Z' column using the estimates of the main effects obtained from the solution of the matrix equation, but for most purposes, unless electronic computing facilities are available, one iteration will suffice.

Testing the adequacy of the model. In the model just used it has been assumed that if any interactions exist between the symptoms these are insignificant. For this reason main effects only have been estimated, and as a result it is desirable to have a test of how well the model fits the data. Dyke and Patterson have supplied a chi-square test of goodness of fit for this purpose. To apply it a new Z column, which will be denoted Z'', is set up. The entries in it are found from the estimates m', d', a' and g' in the same way that the entries Z' were found from the original estimates of the main effects. For example the fourth entry in the Z''-column, given below, is:

$$m' - d' + a' + g'$$

corresponding to the symbol ag.

The Z'' values in turn are converted into R-values using Fisher and Yates's Table VII. These values are given below:

TABLE 6.6

Testing for Goodness of Fit

Treatment	Z''	R''	$n(r-R'')^2/(1-R''^2)$
(l)	$-0\cdot3244$	$-0\cdot3135$	$0\cdot1205$
g	$-0\cdot0744$	$-0\cdot0743$	$0\cdot0333$
a	$-0\cdot1343$	$-0\cdot1335$	$0\cdot2059$
ag	$0\cdot1158$	$0\cdot1153$	$0\cdot0808$
d	$0\cdot0080$	$0\cdot0080$	$0\cdot0000$
dg	$0\cdot2580$	$0\cdot2524$	$0\cdot7067$
da	$0\cdot1981$	$0\cdot1956$	$0\cdot0000$
dag	$0\cdot4482$	$0\cdot4204$	$0\cdot2444$
		Total	$1\cdot3916$

The chi-square test is given by the formula:

$$\chi^2 = \sum \frac{n(r-R'')^2}{1-R''^2}$$

The required components are given in the final column of Table 6.6 and the value of χ^2 is found to be $1\cdot392$. The degrees of freedom are $2^n - k$, where n is the number of symptoms and k is the number of parameters being estimated. For our data n is 3 and k is 4, so that the number of degrees of freedom is 4. Referring the value $1\cdot392$ to the χ^2 tables with four degrees of freedom it is found not to be significant. It is therefore safe to assume that our model which takes main effects into account but neglects interaction effects is adequate for accounting for the variation in the data. However, if in a given situation we were interested in possible interaction effects then the calculations would require adjustment so that these effects could be estimated and examined. Dyke and Patterson in their article indicate how to proceed under such circumstances, but, as might be expected, the calculations now become even more arduous. If all the interactions are taken into account then the matrix on the left of our matrix equation becomes an 8×8 matrix and it is in inverting it that most of the labour lies. Of course all the interaction effects may not be considered of equal

importance and the work can be reduced by omitting those in which we are least interested. To set up the 8×8 matrix a symbolic matrix of these dimensions is first set up by multiplying a column vector and a row vector, each consisting of the symbols 1, g, a, ag, etc., together. In the latter matrix squares such as a^2, b^2, etc., are then replaced by unity while all the diagonal entries become (1); finally the symbols are replaced by the numerical values representing them in column (5), Table 6.5.

REFERENCES

BARTLETT, M. S. (1935) 'Contingency table interactions', *Journal Royal Statistical Society*, (Supplement), **2**, 248–52

DWYER, P. S. (1952) '*Linear Computations*', New York, Wiley & Sons.

DYKE, G. V. & PATTERSON, H. D. (1952) 'Analysis of factorial arrangements when the data are proportions', *Biometrics*, **8**, 1–12

FISHER, R. A. & YATES, F. (1957) *Statistical Tables for Biological, Agricultural and Medical Research*, Edinburgh, Oliver & Boyd

FREEMAN, G. W. & HALTON, J. H. (1951) 'Note on an exact treatment of contingency, goodness of fit and other problems of significance', *Biometrika*, **38**, 141–9

KENDALL, M. G. (1949) 'Rank and product moment correlation', *Biometrika*, **36**, 177–93

MATHER, K. (1949) *Statistical Analysis in Biology*, 3rd ed., London, Methuen & Co. Ltd.

MAXWELL, A. E. (1958) *Experimental Design in Psychology and the Medical Sciences*, London, Methuen & Co. Ltd.

THOMSON, G. H. (1951) *The Factorial Analysis of Human Ability*, London, London University Press

WINSOR, C. P. (1948) 'Factorial analysis of a multiple dichotomy', *Human Biology*, **20**, 195–204

YATES, F. (1937) *The Design and Analysis of Factorial Experiments*. Harpenden, Imperial Bureau of Soil Science

CHAPTER VII

Tests of Goodness of Fit

Introduction. So far the chi-square test has been considered only as a test of 'association', but it has important uses too as a test of 'goodness of fit'. These occur when a frequency curve is fitted to data and it is desired to test whether the discrepancies between the observed frequencies and those derived from the theoretical distribution being fitted are greater than could reasonably be attributed to chance fluctuations in the observed data. When working with frequency data the theoretical distributions which are commonly found useful are the binomial, the Poisson and the negative binomial. But apart from mathematical distributions such as these it is regularly the case that data obtained by sampling have been classified into categories for which the expected frequencies for some related population are known and it is desired to test whether the sample may be considered to have been drawn from that population. Such cases will be considered first; later in the chapter a discussion of the problems of testing goodness of fit when theoretical distributions are concerned will be given. It will be shown too how the customary chi-square test can often be improved upon by the use of 'variance tests', and by tests of linear functions of deviations between observed and expected values (Cochran, 1955).

Goodness of Fit Tests when the Expected Frequencies are Known

To illustrate the procedure we will consider some data from the *Triennial Report* (1949–51) of the Bethlem Royal and Maudsley Hospitals, by Blacker and Gore. They noted that the fertility of hospital patients, as measured by the number of children they had, differed somewhat from the figures for the population of England and Wales as a whole. For 2565 female patients the frequency distribution for

different sizes of family are shown in Table 7.1, column (2). In column (3) these frequencies have been expressed as percentages of the total sample size, while in column (4) of the same table corresponding percentages for the population are given. When the latter percentages are multiplied by the sample size, namely 2565, and divided by 100 the number of patients one would expect for each size of family, were the distributions for patients and population the same, are obtained. The problem now is to test whether the discrepancies $(O-E)$ between ob-

TABLE 7.1

Frequency Distribution of Married Female Patients for Different Sizes of Family together with Expected Frequencies Based on the Normal Population

(1) Family size	(2) *Observed number (O) of patients*	(3) *Percentage number of patients*	(4) *Percentage expected*	(5) *Number (E) expected*	(6) *Discrepancies (O − E)*	(7) *Difference of col. (3) − (4)*
0	612	23·9	21·6	554·0	58·0	2·3
1	769	30·0	30·3	777·2	− 8·2	−0·3
2	643	25·0	26·5	679·7	−36·7	−1·5
3	272	10·6	11·8	302·7	−30·7	−1·2
4	112	4·4	5·0	128·3	−16·3	−0·6
5–6	103	4·0	3·3	84·6	18·4	0·7
7–9	41	1·6	1·3	33·3	7·7	0·3
>9	13	0·5	0·2	5·1	7·9	0·3
Total	2565	100·0	100·0	2565	0	0·0

served (O) and expected (E) frequencies are indicative of a real difference between patients and population. (The comparison, as Blacker and Gore point out, is not quite exact in as far as the national figures exclude women over fifty, while the hospital figures include 500 or so women above that age. However, it is known that the fertility of these women is higher than that of the younger women so the discrepancy between observed and expected frequencies in Table 7.1 is probably an under-estimate.)

To test whether the discrepancies $(O-E)$ are greater than could reasonably be attributed to chance causes a chi-square test is performed in the ordinary way. For each category the quantity $(O-E)^2/E$

is calculated. The sum of these quantities is the value of chi square required. It is referred to the chi-square table with $(k-1)$ degrees of freedom, where k is the number of categories in the classification distribution.

For the data in Table 7.1 chi square is found to be $31 \cdot 344$. Now the data are classified into eight categories but once the sample size is fixed only seven of these can be filled arbitrarily so that there are only seven degrees of freedom. Referring $31 \cdot 344$ to the chi-square table with seven degrees of freedom we find the probability of so large a value having arisen by chance to be less than $0 \cdot 001$, so it is safe to conclude that there is a real difference between the observed frequencies for our sample of patients and those that would be expected were the sample drawn randomly from the population as a whole. The nature of this difference will be discussed later.

Lower limits for expected frequencies. In Table 7.1 all the expected frequencies are greater than 5 so that the validity of the chi-square test is not called into question. Had they not been, then, following normal procedure, two or more adjacent categories would be combined until all the expected frequencies were brought up to an acceptable level. In the data in our example a certain amount of telescoping of categories had already been done before the data were published so that further adjustment was unnecessary. However, the rule that no expected frequency should fall below 5 is not sacrosanct, and though enough is not yet known to fix the lower limit satisfactorily there is some evidence to show (Cochran, 1954) that when the number of categories is five or so at least one expected frequency as low as unity has negligible effect on the significance level of the test and may safely be admitted. This point is worth emphasising for it is often the case that the main difference between an observed and an expected distribution is found to lie at the tails of the distribution and might be overlooked or blurred if categories were combined unnecessarily.

Testing a Linear Function of the Deviations

To appreciate where the discrepancies between patients and normals lie in the example considered it is well to look at the entries in column (7) of Table 7.1, where the differences between the percentages in the

different categories for patients and normals are shown. There it is seen that there is an excess of 2·3 per cent of patients over normals in the category 'no-children', while in the categories 'two children' and 'three children' patients fall noticeably below normals to the extent of 1·5 per cent and 1·2 per cent respectively.

When dealing with samples much smaller than that used in our example an ordinary overall χ^2 test may not be sufficiently sensitive to detect discrepancies of the magnitude mentioned. This is so because the overall chi-square test is not directed against any specific pattern of deviations. However, when it is possible to state in advance the direction and relative magnitude of certain discrepancies Cochran (1955) has provided an alternative to the chi-square test which is more sensitive for detecting a failure of the null hypothesis.

To do the test a set of weights g_i, where i refers to categories, are chosen *in advance*, that is on prior knowledge, by the experimenter in such a way that the linear expression:

$$L = \sum g_i(O_i - E_i) \qquad \ldots 7.1$$

will be sensitive to discrepancies of the type predicted. For instance a possible set of weights to use in a future study for detecting differences between the fertility of mental patients and the normal population would be those given in column (7) of Table 7.1, the algebraic signs of the percentage differences given in that column being retained. Alternatively, the weights could be simplified to read:

$$2, \quad 0, \quad -1, \quad -1, \quad 0, \quad 0, \quad 0, \quad 0,$$

for the eight categories respectively, where differences of less than 1 per cent are ignored.

Using the latter weights and a fresh sample Cochran's test can be employed to see if the difference between the fertility—as indicated by family size—of patients and normal controls is confirmed. A comparatively small sample of just 200 patients is used. The classification of these patients into categories depending on the size of the families they have is given in column (3) of Table 7.2. In column (4) of the same table the expected frequencies for the categories for a sample of 200 are shown. These were calculated using the percentages in column (4) of Table 7.1. As the expected frequency for the last

category (viz. more than nine children) is less than unity this category can be combined with that preceding it.

TABLE 7.2

Frequency Distribution for a Sample of 200 Female Patients

Category No. i	Family size	Observed frequency (O_i)	Expected frequency (E_i)	Weights g_i	$O_i - E_i$	$g_i E_i$	$g_i^2 E_i$	$g_i(O_i - E_i)$
0	0	56	43·2	2	12·8	86·4	172·8	25·6
1	1	58	60·6	0	− 2·6			
2	2	43	53·0	−1	−10·0	−53·0	53·0	10·0
3	3	20	23·6	−1	− 3·6	−23·6	23·6	3·6
4	4	12	10·0	0	2·0			
5	5–6	6	6·6	0	− 0·6			
6	7–9	2 }5	2·6 }3	0	2·0			
7	>9	3	0·4					
Total		200	200·0		0·0	9·8	249·4	39·2

In the special case where the expected frequencies are known *a priori* and so have not to be calculated from the data, as incidentally is so for the above data, Cochran has shown that an estimate $V(L')$ of the variance of L—equation 7.1—is given by the expression:

$$V(L') = \sum g_i^2 E_i - (\sum g_i E_i)^2 / N \qquad \ldots 7.2$$

and that when N, the sample size, is fairly large L' tends to be normally distributed so that a test of the significance of the null hypothesis—which in our example is that patients do not differ from normals—can be made by referring the ratio:

$$L/\sqrt{[V(L')]}$$

to the normal curve.

The data required for substitution in equations 7.1 and 7.2 to test the null hypothesis are tabulated in Table 7.2. On making the substitutions it is found that:

$$L' = 39·2$$

while $$V(L') = 249·4 - 9·8^2/200 = 248·92$$

The estimate of the standard error of L' is therefore $\sqrt{(248·92)} = 15·78$, so that the critical ratio is $39·2/15·78 = 2·48$. The probability of a value as large or larger than this is less than $0·02$ so that the null hypothesis may safely be discarded. Here it must be emphasised that if a chi-square test is performed in the ordinary way on the discrepancies between observed and expected frequencies in Table 7.2—the last two categories being combined—a value of χ^2 equal to $8 \cdot 129$ is obtained. With six degrees of freedom this value does not reach even the $0 \cdot 2$ probability level so that the greater sensitivity of Cochran's test as compared with an overall chi-square test is clearly demonstrated.

When Expected Frequencies are Estimated from the Data

In the last example the expected frequencies for the several categories concerned were known for the population. Frequently such is not the case and they have to be estimated from the sample data. For instance in the paper by Cochran (1955) already referred to a study is reported in which a binomial distribution is fitted to data for 53,680 families of exactly eight children classified into the nine categories in which the number of boys in the family was 0, 1, 2, ... 7, 8, respectively. Here the expected frequency in categories 0 to 8 is given by multiplying the successive terms in the expansion of the binomial expression:

$$(q+p)^8$$

by N, the sample size. In this expression p is the proportion of boys in families of size eight in the population, and $p+q = 1$. As this proportion is not known an estimate of it has to be obtained from the sample. It is known that an efficient estimate of p, when the distribution is assumed to be binomial, is the sample mean given by the formula $\sum(iO_i)/N$, where i refers to categories and in the example in question runs from 0 to 8, and O_i is the observed frequency in the i-th category: N once again is the sample size.

The expansion of $(q+p)^n$, when n is a positive whole number, has $(n+1)$ terms. They are the respective additive terms in:

$$(q+p)^n = q^n + nq^{n-1}p + \frac{n(n-1)}{2!}q^{n-2}p^2 + \frac{n(n-1)(n-2)}{3!}q^{n-3}p^3 +$$
$$\dots + nqp^{n-1} + p^n \qquad \dots 7.3$$

where $k! = k(k-1)(k-2) \dots 3 \times 2 \times 1$. When n is taken as 8 the first term in the expansion, q^n, is the probability that a family of eight drawn at random from the population concerned will be all girls. The second term, $nq^{n-1}p$, gives the probability that a family drawn at random will have seven girls and one boy; the third term gives the probability of drawing a family with six girls and two boys; and so on for the other terms. When these probabilities are multiplied in turn by the sample size, namely 53,680, for the case in point the expected relative frequencies of families of eight consisting of different numbers of boys and girls for a sample of the size drawn are obtained. The observed frequencies in the several categories can now be compared with the expected frequencies by a chi-square test, as shown in the last example, to see whether the assumption that the data are distributed according to the binomial distribution is tenable. Linear functions of the discrepancies between observation and expectation can also be considered by the methods outlined in the previous section.

Dispersion Tests for Binomial Distributions

In cases where it is suspected that the observed frequencies—though ideally they might be expected to follow a binomial distribution—depart from it because:

(a) the probability p is subject to some kind of random variation from observation to observation, or the variance of the observed distribution is larger than that of the theoretical distribution;

(b) the probability p is affected by a systematic source of variation as a series of observations is taken;

a more sensitive comparison than that provided by the 'goodness of fit' test can often be obtained by comparing the variances of the observed and the expected distributions. The relevant tests are referred to as *dispersion tests* or *variance tests*.

In the case of the binomial distribution Cochran (1954) lists four tests of this kind to meet situations in which:

(i) n is constant and p is given in advance;

(ii) n is constant and p is estimated from the data;

(iii) n varies and p is given in advance;

(iv) n varies and p is estimated from the data.

These variance tests, first introduced by Fisher, are also appropriate in cases in which the number of observations is too small to admit the use of 'goodness of fit' tests. Situation (ii) above is perhaps the one which occurs most frequently in practice and the variance test required in that situation will now be illustrated.

In a pursuit rotor experiment a subject was required to touch with a stilus a silver spot placed off-centre on a black revolving disk (on a gramophone table). The disk revolved once a second and the subject was required to make one attempt to touch the silver spot during each

TABLE 7.3

Pursuit Rotor Data

Score i	Observed frequencies O_i	Expected frequencies E_i	$(i - n\hat{p})$	$(i - n\hat{p})^2$	$(O_i - E_i)$	$\dfrac{(O_i - E_i)^2}{E_i}$
0	41	43·23	−1·4609	2·134	−2·23	0·1150
1	86	89·14	−0·4609	0·212	−3·14	0·1106
2	82	73·53	0·5391	0·291	8·47	0·9757
3	31	30·32	1·5391	2·369	0·68	0·0153
4	3	6·25	2·5391	6·447	−3·77	2·0994
5	0	0·52	3·5391	12·525		
	243	242·99				$x^2 = 3\cdot3160$

$\sum iO_i = 355$; $\sum O_i(i - n\hat{p})^2 = 222\cdot368$; $\bar{x} = n\hat{p} = 355/243 = 1\cdot4609$.

revolution of the disk. Contact with the spot was indicated by the sound of a bell connected to the apparatus. Attempts were made by the subject in successive runs of five followed by a short rest period. The score in each run was the number of successes out of 5. The distribution of successes for 243 runs is given in Table 7.3.

In the same table the expected frequencies, assuming a binomial distribution, are given. To obtain these an estimate $\hat{p}$ of p, the probability of success on a single attempt, was obtained from the data. It is given by the formula:

$$n\hat{p} = \bar{x} \qquad \qquad \ldots 7.4$$

where $\bar{x} = (\sum iO_i)/N$ is the mean of the distribution of scores. Since $\sum iO_i = 355$, $N = 243$ and $n = 5$, $\hat{p}$ is found to be $0\cdot292$. The expected frequencies (Table 7.3) are now given by the six terms in the expansion of:

$$N(\hat{q} + \hat{p})^5$$

where $\hat{q} = (1 - \hat{p}) = 0.708$.

The problem is to test whether the discrepancies between the observed frequencies and those given when a binomial distribution is assumed are greater than could reasonably be attributed to chance. The expected frequencies correspond to the situation in which p, the probability of success at any attempt, remains constant throughout the experiment. But this assumption may not be justified for the subject may improve with practice, or his performance may deteriorate due to boredom or fatigue, or be otherwise affected by chance factors. For these reasons it seems more appropriate to test the discrepancies between the observed and the theoretical distributions by a variance test than by the χ^2 test of goodness of fit.

The variance test. The relevant variance test in the situation in which n is fixed and p is estimated from the data requires us to evaluate the expression:

$$\chi_v^2 = \frac{\sum O_i(i - n\hat{p})^2}{n\hat{p}\hat{q}} \qquad \ldots 7.5$$

and to refer the result to the χ^2 table with $(N-1)$ degrees of freedom, where N is the sum of the observed frequencies. In equation 7.5, $\hat{p}$ is the estimated value of the probability of a success at a single trial and was found above to be 0.292, while $\hat{q}$ is $1 - 0.292 = 0.708$. From Table 7.3 $\sum O_i(i - np)^2$ is found to be 222.368 so that:

$$\chi_v^2 = 222.368/(5 \times 0.292 \times 0.708) = 215.12$$

Since the number of degrees of freedom, $N-1 = 242$, is large the significance of the obtained value of χ^2 is tested by referring the value of the expression $\sqrt{(2\chi^2)} - \sqrt{(2 \times 242 - 1)}$ to the normal distribution. The value of this expression is -1.23, which corresponds to a probability of $P = 0.109$, so that the discrepancies between observed and expected frequencies are not significant.

Had a χ^2 test of goodness of fit been applied to the data in the ordinary way, the last two expected frequencies being combined so that no expected frequency was less than unity, the contribution of each discrepancy to the total χ^2 would be the corresponding value of $(O_i - E_i)^2/E_i$ shown in the last column of Table 7.3. The overall value of χ^2 in this case is $3 \cdot 316$. Moreover, since there are only five categories after the frequencies for scores four and five have been combined, and since two degrees of freedom are used up in fixing the sample size and in estimating p, the obtained value of χ^2 of $3 \cdot 316$ is based on $5 - 2 = 3$ degrees of freedom. This value, when referred to the chi-square table, corresponds to a probability of $P = 0 \cdot 29$, which is considerably larger than the probability $P = 0 \cdot 109$ obtained from the variance test, so that the greater sensitivity of the latter test is demonstrated. In conclusion it is of interest to note that the greatest contribution to the overall χ^2 value in the last column of Table 7.3 comes from the combined category for scores 4 and 5. Here the observed frequency is less than the expected frequency by $3 \cdot 77$, suggesting that the number of occasions on which the subject got a high score in a 'run' of five attempts is less than one would predict on the basis of the binomial distribution.

The Poisson Distribution

Associated with the binomial distribution is the Poisson distribution. It can be obtained from the binomial by allowing the value of p in the expression $(q+p)^n$ to become very small and at the same time increasing n sufficiently for np to be finite. The distribution is of value in situations, of frequent occurrence in demographic work, in which the probability of an event happening is very small but the number of occasions on which the event could happen is so large that it is observed a considerable number of times.

In describing the Poisson distribution reference must be made to the mathematical constant e, which incidentally is the base of natural or Naperian logarithms. Its value, which correct to three decimal places is $2 \cdot 718$, is given by the sum of the terms to infinity of the series:

$$e = 1 + 1/1! + 1/2! + 1/3! + \ldots + 1/k! + \ldots$$

where, as before $k! = k(k-1)(k-2)\ldots 3 \times 2 \times 1$. If e is raised to the power m the series becomes:

$$e^m = 1 + m + m^2/2! + m^3/3! + \ldots + m^k/k! + \ldots$$

The terms of the Poisson series, in which we are interested, are given by multiplying each term of the last expression by e^{-m}; they are given by:

$$e^{-m}(1+m+m^2/2!+m^3/3!+m^4/4!\ldots \qquad \ldots 7.6$$

Since $e^{-m} \times e^m = 1$ the value of this expression is unity.

From the statistical viewpoint the usefulness of the Poisson series lies in the fact that the sum of its terms is unity, just as the sum of the terms in the expansion of the binomial expression $(q+p)^n$ is unity when $(q+p)$ is taken equal to 1. This permits it to be used as a probability distribution for in a situation in which the terms of a series stand for the probability of an event falling into one or other of a set of mutually exclusive categories the sum of these probabilities must be unity: in other words, the categories must be exhaustive. To clarify the matter the process by which a Poisson distribution is fitted to experimental data will be illustrated by an example. In addition, tests for assessing the goodness of fit will be discussed.

Fitting the Poisson distribution. Distributions of the Poisson type are found frequently in work with morbidity data. For example the distribution given in Table 7.4 was obtained by L. Stein and others

TABLE 7.4

Distribution of Spells of Sickness Absence for 892 Females

Number of spells i	Frequency of spells O_i	Expected frequency E_i	Product iO_i	Discrepancy $(O_i - E_i)$	Chi square $(O_i - E_i)^2/E$
0	537	494·6	0	42·4	3·64
1	229	291·7	229	−62·7	13·48
2	95	86·0	190	9·0	0·94
3	19	16·9	57	2·1	0·26
4	10 ⎱	2·8	40 ⎱	9·2	30·23
5 +	2 ⎰		10 ⎰		
Totals	892	892·0	526	0·0	48·55 = x^2

when sampling the *spells* of sickness absence amongst local authority employees in Edinburgh for the period June 1952 to May 1954. By a spell of sickness absence they meant a period of four days or more; in other words, they were concerned with 'insured sickness spells'. The distribution is of the J-shaped type typical of Poisson distributions, but the student should note that the maximum frequency or mode of such distributions does not necessarily always occur at the left-hand tail—the curve may have a turning point to the right of this tail.

From the data in Table 7.4 it is seen that of the sample of 892 females considered 537 had no spell of sickness lasting as long as four days, 229 had one spell of sickness absence of four days or more, 95 had two spells of four days or more, etc. The problem now is to see how well this observed distribution follows one of the Poisson type.

In the Poisson expansion (expression 7.6) m stands for the mean of the distribution—in the present example the average number of spells of sickness. Since m is not known in advance it must be estimated from the data in the sample. An efficient estimate of it is given by the formula:

$$m = \frac{\sum iO_i}{N} \qquad \ldots 7.7$$

where i is the category number (i.e. the number of spells) and O_i is the frequency in the i-th category; N is the sample size.

For the data in Table 7.4 m is found to be $526/892 = 0 \cdot 590$ spells per individual. The expression e^{-m}, namely $2 \cdot 71828^{-0 \cdot 59}$ is now evaluated. This is done as follows:

Let $\qquad\qquad\qquad k = 2 \cdot 71828^{-0 \cdot 59}$

then $\qquad\qquad\quad \log k = -0 \cdot 59 \times \log 2 \cdot 71828$

$\qquad\qquad\qquad\qquad = -0 \cdot 59 \times 0 \cdot 43429$

$\qquad\qquad\qquad\qquad = -0 \cdot 25609 \quad\text{or}\quad \bar{1} \cdot 74391$

hence $\qquad$ antilog $k = 0 \cdot 55451 = e^{-m}$

Substituting this value, and the value $m = 0 \cdot 590$, in expression 7.6 we obtain the series:

$$0 \cdot 55451(1 + 0 \cdot 59 + 0 \cdot 3477/2 + 0 \cdot 2051/6 + 0 \cdot 1209/24 + \ldots)$$

On evaluating the first five terms of the series the probabilities $0 \cdot 55451 + 0 \cdot 32699 + 0 \cdot 09641 + 0 \cdot 01895 + 0 \cdot 00279 + (0 \cdot 00035)$ are obtained. The sixth term in the series is a composite term corresponding to the sixth category in Table 7.4. It is obtained by subtracting the sum of the first five terms from unity. These probabilities are now multiplied by 892, the sample size, to obtain the frequencies one would expect if the observed distribution really follows the Poisson law. The expected frequencies are given in column three of Table 7.4, but as the last two, $2 \cdot 5$ and $0 \cdot 3$, are well below 5 they have been combined with a view to testing the goodness of fit of the distribution.

Testing the goodness of fit of the Poisson distribution. A chi-square test can now be performed on the data in columns two and three, Table 7.4. The discrepancies between the observed (O_i) and the expected (E_i) frequencies are given in column five of the table and chi square is calculated in the usual way using the formula

$$\sum (O-E)^2/E$$

The value obtained is $48 \cdot 55$. The degrees of freedom to be attached to this value need careful attention. In obtaining it five categories were employed, for though the expected frequency for one of these categories is less than 5 this category was retained for reasons discussed earlier. With five categories four degrees of freedom are available. But in fitting the Poisson distribution the parameter m had to be estimated from the data themselves and this utilises a further degree of freedom. Hence the value $48 \cdot 55$ for chi square is based on three degrees of freedom and incidentally is highly significant. The general rule about degrees of freedom when a theoretical distribution is fitted to frequency data is that one degree of freedom is lost for each parameter which has to be estimated from the data; in addition one degree of freedom is used up in fixing the sample size, for if there are n cells or categories only $(n-1)$ of them can be filled arbitrarily.

As we have seen the observed frequencies in Table 7.4 differ very significantly from those which would be expected had the data followed a Poisson distribution. Examination of the entries in columns five and six of that table shows that the biggest discrepancies are in categories 0, 1 and 5. Here it should be noted that in years when an epidemic, such as an influenza epidemic, occurs the observed distribution shows fewer entries in category 0 and more in category 1. In such cases the observed frequency distributions follow the Poisson distribution much more closely. When this is the case a more exact test of goodness of fit, than that afforded by an ordinary chi-square test, can be made—as Cochran (1955) has shown—if a linear function, $L = \sum g_i(O_i - E_i)$, of the discrepancies can be set up. The weights g_i require to be chosen in advance and are based on prior knowledge of the relative magnitudes and directions of the discrepancies to be expected. From our examination of the data in Table 7.4 we might choose the weights 1, −1, 0, 0, and 2 for the successive categories in that investigation. With these in mind let us look at another frequency distribution, Table 7.5, obtained in an investigation similar to that discussed above.

TABLE 7.5

A Further Distribution of Spells of Sickness Absence

(1) i	(2) g_i	(3) O_i	(4) E_i	(5) $g_i(O_i - E_i)$	(6) $g_i E_i$	(7) $g_i^2 E_i$	(8) $(i - m)$	(9) $g_i E_i(i - m)$
0	1	175	166·5	8·5	166·5	166·5	−0·52	−86·580
1	−1	72	86·5	14·5	−86·5	86·5	0·48	−41·520
2	0	27	22·5	0·0	0·0	0·0	1·48	0·0
3	0	4	3·9	0·0	0·0	0·0	2·48	0·0
4 +	2	2	0·6	2·8	1·2	2·4	3·48	4·176
Total		280	280·0	25·8	81·2	255·4		−123·924

For this distribution the mean number of spells of illness is estimated to be 0·52. The observed frequencies appear in column (3) and the estimated expected frequencies, assuming a Poisson distribution, are given in column (4). A chi-square test performed on the entries in these columns, combining the last two categories, gives a value $\chi^2 = 4·36$. With two degrees of freedom this value falls short of the 5 per cent significance level so that we might conclude that the data followed a Poisson distribution fairly well.

However, using the weights given in column (2) of the table Cochran's more sensitive test can be carried out. The data required are tabulated in Table 7.5. The linear function:

$$L = \sum g_i(O_i - E_i)$$

has the value 25·8. An estimate of the variance of L has now to be obtained. In the present case the expression given in equation 7.2 is not appropriate since the mean m, and consequently the expected frequencies employed in this example, had to be estimated from the sample itself. In such cases Cochran has shown that an estimate of the variance of L is given approximately by the expression:

$$V(L) = \sum g_i^2 E_i - (\sum g_i E_i)^2/N - [\sum g_i E_i(i-m)]^2/(Ns^2) \ldots 7.8$$

In this expression m and s^2 are estimates of the mean and variance respectively of the distribution. However, for the Poisson distribution it is known that $s^2 = m$, so that for our example the required estimates are each equal to $0 \cdot 52$. (Had the Binomial distribution, rather than the Poisson, been concerned then having estimated m as above an estimate of s^2 could be obtained by the formula $s^2 = m(1 - m/N)$.

Using the data from Table 7.5 we find:

$$V(L) = 255 \cdot 4 - 81 \cdot 2^2/280 - (-123 \cdot 9)^2/(280 \times 0 \cdot 52)$$
$$= 126 \cdot 4$$

To obtain the test criterion we calculate:

$$L^2/V(L)$$

It is $665 \cdot 6/126 \cdot 4 = 5 \cdot 27$, and is referred to the chi-square table with one degree of freedom. The result is significant almost at the 2 per cent level. When we recall that a chi-square test applied to the data in Table 7.5 in the ordinary way did not prove to be significant the greater sensitivity of Cochran's linear test is demonstrated.

Testing the significance of a single deviation $(O_i - E_i)$. In addition to testing a linear function of the deviations between observed and expected frequencies any single deviation $(O_i - E_i)$ can be tested for significance provided the category to which it refers is chosen in advance. To do the test the quantities required are:

$$L = (O_i - E_i) \qquad \ldots 7.9$$

and an estimate of the variance of L given by the expression:

$$\hat{V}(L) = E_i - \frac{E_i^2}{N}\left\{1 + \frac{(i-m)^2}{m}\right\} \qquad \ldots 7.10$$

The value of χ^2 given by the ratio,

$$\chi^2 = L^2/V(L)$$

is then referred to the chi-square table with one degree of freedom.

The expression for the estimate of the variance of L given above is in the form required when a Poisson distribution is under consideration. When the distribution is binomial m is replaced by $n\hat{p}$.

Variance tests for the Poisson distribution. As in the binomial case departure from the Poisson distribution frequently occurs because the mean of the distribution itself varies from group of observations to group of observations or even from one observation to another. It may vary too in a systematic way as the observations are made. In such cases a comparison between the observed variance of the distribution and the variance predicted from Poisson theory is often more sensitive for detecting real discrepancies than a goodness of fit test.

When the mean of the distribution has to be estimated from the data the variance test is made by calculating:

$$\chi_v^2 = \sum \frac{O_i(i-m)^2}{m}$$

where m is the estimate of the mean of the distribution, and referring the result to the chi-square table with $N-1$ degrees of freedom, N being the sum of the observed frequencies. Other related variance tests, for which Cochran (1954) gives appropriate test-statistics, enable one to examine the data for:

(a) an abrupt change in the mean of the distribution;
(b) a linear trend in the mean;
(c) a point at which a change in level occurs;

but space does not allow these tests to be considered in detail here.

In concluding this chapter on goodness of fit tests it is well to say that though the negative binomial distribution—which was mentioned in passing—will not be discussed there are two papers in particular, one by Anscombe (1950) and the other by Evans (1953), which the student wishing to use it will find helpful. An interesting example in which the curve is used to describe the pattern of consumer purchases has been reported by Ehrenberg (1959).

REFERENCES

ANSCOMBE, F. J. (1950) 'Sampling theory of the negative binomial and logarithmic series distributions', *Biometrika*, **37**, 358–82

BLACKER, C. P. & GORE, A. T. (1952) *Triennial Statistical Report* (1949–1951), Bethlem Royal and Maudsley Hospitals

COCHRAN, W. G. (1955) 'A test of linear function of the deviations between observed and expected numbers', *Journal of the American Statistical Association*, **50**, 377–97

COCHRAN, W. G. (1954) 'Some methods for strengthening the common χ^2 tests', *Biometrics*, **10**, 417–51

EHRENBERG, A. S. C. (1959) 'The pattern of consumer purchases', *Applied Statistics*, **8**, 26–41

EVANS, D. A. (1953) 'Experimental evidence concerning contagious distributions in ecology', *Biometrika*, **40**, 186–211

KENDALL, M. G. (1952) *The Advanced Theory of Statistics*, London, Charles Griffin & Co.

CHAPTER VIII

Rank Correlation Techniques

Introduction. Ranking methods are sometimes thought to be the *sine qua non* of statistical techniques for psychologists and social scientists. Despite this ranking methods are not widely used in these fields—except perhaps in inefficient and inappropriate ways, say when subjects are ranked on the basis of quantitative measures made on them and a rank correlation coefficient is substituted for the product-moment coefficient to reduce the amount of calculations to be performed. Ranking methods are seldom used primarily because of the unsatisfactory nature of the ranking routine. Few teachers, for instance, would try—except under duress—to rank a large class of children with regard to their ability at or aptitude for a subject such as mathematics, much less subjects such as Music or Art. They would always resort to some kind of test or questionnaire from which quantitative measures could be obtained. Nevertheless if presented with the children in small sub-groups, three or four at a time, most teachers would be prepared to rank them; though were the sub-groups of minimum size—two, as in paired comparisons—the number of possible comparisons would now be so great, if the class were large, that the task would become extremely tedious should one person be required to make all the comparisons. Practical considerations of this kind must be taken into account when planning an investigation in which it is proposed to use ranking methods. These considerations will, to a large extent, influence the choice of topics to be discussed here. In particular special attention will be paid to the use of incomplete blocks in ranking experiments (Durbin, 1951) which allows the material being ranked to be split up into tractable sub-groups. However, some mention of the more elementary ranking methods will first be made.

Spearman's Rank Correlation Coefficient—Rho (ρ)

This coefficient is the oldest and best known of the rank correlation coefficients. It is a measure of the correlation between two rankings and is given by the formula:

$$\rho = 1 - \frac{6 \sum d_i^2}{n(n^2 - 1)} \qquad \ldots 8.1$$

where d_i ($i = 1, 2, \ldots, n$) is the difference between the two rankings for the i-th person or thing being ranked and n is the sample size. Rho ranges in value from $+1$, when the rankings on the two occasions are identical, to -1, when the rankings are as different as possible, that is when the person or thing ranked highest on one occasion is ranked lowest on the other, the second highest second lowest, and so on.

Formula 8.1 can be obtained directly from the formula for the product moment correlation coefficient between two continuous variables X and Y, namely:

$$r_{xy} = \frac{\sum xy}{\sqrt{(\sum x^2 . \sum y^2)}} = \frac{\sum x^2 + \sum y^2 - \sum (x-y)^2}{2\sqrt{(\sum x^2 . \sum y^2)}}$$

(for details see pp. 65 *et seq.*) by putting $\sum(x-y)^2 = \sum d^2$, and by substituting:

$$\frac{\sum x^2}{n} = \frac{\sum y^2}{n} = \frac{(n^2 - 1)}{12}$$

$(n^2 - 1)/12$ being the variance of the first n natural numbers.

For $n \geqslant 10$ the significance of rho can be tested (approximately) by referring:

$$t = \rho\sqrt{[(n-2)/(1-\rho^2)]} \qquad \ldots 8.2$$

to the t-distribution with $n-2$ degrees of freedom. For values of n from 5 to 10 the values which rho must reach to be significant at the 5 per cent level are (approximately) as given in Table 8.1.

TABLE 8.1

5 per cent Significance Values of Rho for n from 5 to 10

n	5	6	7	8	9	10
ρ	0·96	0·90	0·83	0·76	0·71	0·68

An example. To illustrate the use of *rho* consider the following example. Students training as clinical psychologists are ranked by a tutor at the end of their course as regards (*a*) their suitability for their career, and (*b*) their knowledge of psychology (alternatively, they might be ranked by two tutors on one or other of (*a*) and (*b*)), and it is desired to test whether there is a significant correlation between their rankings. In a particular year there are 10 students ($n = 10$), $A, B, \ldots J$, who—it is felt—are a representative sample of the type of student met on the course. They are ranked on *a* and *b* their rankings from 1, for the best student, to 10, for the worst student, being:

Student:	A	B	C	D	E	F	G	H	I	J
Ranking on *a*	4	10	3	1	9	2	6	7	8	5
Ranking on *b*	5	8	6	2	10	3	9	4	7	1

The difference (*d*) and the difference squared (d^2) between the respective pairs of ranks are:

d	−1	2	−3	−1	−1	−1	−3	3	1	4
and d^2	1	4	9	1	1	1	9	9	1	16 Total = 52

$\sum d^2$ is therefore 52, so that on substitution in equation 8.1 the value of *rho* is found to be:

$$\rho = 1 - (6 \times 52)/(10 \times 99)$$
$$= 0 \cdot 737$$

To test this correlation coefficient for significance we substitute $\rho = 0 \cdot 737$ and $n = 10$ in equation 8.2. This gives:

$$t = 0 \cdot 737 \sqrt{\left(\frac{8}{1 - 0 \cdot 737^2} \right)} = 3 \cdot 087$$

When referred to the *t*-table with ($n - 2$), that is 8, degrees of freedom this value of *rho* is seen to be significant beyond the 5 per cent level. This result agrees with the information given in Table 8.1 where it is seen that a value of *rho* equal to $0 \cdot 68$ is required for significance at the 5 per cent level when *n* is 10.

Ties. When ties exist in the rankings the ranks of the cells in which the ties occur are averaged. A simple example will illustrate the point. Suppose that four things, or people, *A* to *D*, are ranked and that *B* and *C* tie for second place, then the ranks 2 and 3 for the second and third places are added and averaged and the average rank, namely $(2+3)/2 = 2 \cdot 5$, is allotted to each of *B* and *C*. If in the complete ranking *A* was first and *D* was last the ranking after adjustment would be:

Item	*A*	*B*	*C*	*D*
Ranking	1	2·5	2·5	4

and the calculations for finding *rho* would proceed as in the earlier example.

Kendall's Rank Correlation Coefficient—Tau (τ)

This coefficient is an alternative to Spearman's *rho* and is preferable to the latter in as far as its sampling distribution is more straightforward. In addition, in the non-null case—that is when the null hypothesis has been discarded—it has an interpretation in terms of the probability that two items selected at random have the same *a* and *b* rankings.

Tau is defined as:

$$\tau = \frac{S}{n(n-1)/2} = \frac{2S}{n(n-1)} \qquad \ldots 8.3$$

where *n* is the number of things ranked and $S = P - Q$, where *P* is the number of pairs having the same rank order in both rankings and *Q* is the number of pairs for which the rank orders disagree. From formula 10.3 it is seen that the coefficient is the difference between *P* and *Q* expressed as a fraction of $n(n-1)/2$, the total number of ways in which the *n* things can be compared two at a time.

To illustrate the calculation of *P* and *Q* the data employed when discussing *rho* will be used. The calculations are simplified if one set of rankings is arranged in natural order as below:

Student:	*D*	*F*	*C*	*A*	*J*	*G*	*H*	*I*	*E*	*B*
Ranking *a*	1	2	3	4	5	6	7	8	9	10
Ranking *b*	2	3	6	5	1	9	4	7	10	8

Now, concentrating on the b-rankings, it is seen that the first rank in the sequence is 2 (corresponding to rank 1 in the a-rankings). It has 8 ranks higher than itself and 1 rank lower than itself to its right so we score 8 for P and 1 for Q. The next rank in the sequence of b-rankings is 3 (corresponding to rank 2 in the a-rankings). It has 7 ranks higher than itself and 1 rank lower than itself to its right, hence we score 7 for P and 1 for Q. Continuing in this way the values which go to make up P and Q are found to be:

$$P = 8+7+4+4+5+1+3+2+0 = 34$$
$$Q = 1+1+3+2+0+3+0+0+1 = 11$$

so that $S = (P-Q) = 34-11 = 23$. On substituting in equation 8.3 *tau* is found to be:

$$\tau = \frac{2 \times 23}{10(10-1)} = 0 \cdot 511$$

Testing the significance of tau. For values of n less than or equal to 10 it can be shown, from tables provided by Kendall (1948), that the values which τ must equal or exceed to be significant at the 5 per cent level of significance are as given in Table 8.2.

TABLE 8.2

The 5 per cent Significance Levels of Tau for n from 5 to 10

n	5	6	7	8	9	10
τ	1·00	0·87	0·71	0·64	0·56	0·51

From this table it is seen that the value of $0 \cdot 511$ for *tau*, obtained from our example, is just significant at the 5 per cent level.

When n is greater than 10 *tau* may be taken to be approximately normally distributed with mean of zero and variance (V) given by

$$V = \frac{4n+10}{9n(n-1)} \qquad \ldots 8.4$$

For our example the value of V is found to be $50/810 = 0 \cdot 0617$, hence the estimate of the standard error of *tau* is the square root of $0 \cdot 0617$, namely $0 \cdot 248$.

When making the test of significance, however, a correction for continuity is essential. This is made by subtracting unity from the absolute value (i.e. the value irrespective of sign) of S before computing *tau*. For the above data the corrected value of S is $(23 - 1) = 22$. The adjusted *tau*, for use in the test of significance, is therefore $(2 \times 22)/(10 \times 9) = 0 \cdot 489$. When this value is divided by $0 \cdot 248$, the estimate of the standard error, the value $1 \cdot 976$ is obtained. This is just above the value $1 \cdot 960$ required for significance at the $0 \cdot 05$ probability level on the normal curve so that the test agrees very closely with the exact test provided by the data in Table 8.2.

When ties occur the calculation of S unfortunately is tricky, and the formula for τ, and more especially that for the variance of τ, are unwieldy and will not be given here. They can be found in Kendall's book (1948). When ties are very few they can be resolved by the toss of a coin and τ can be calculated without bias in the usual way.

A useful summary of the literature on τ and many demonstrations of its use in statistical investigations will be found in an article by Schaeffer and Levitt (1956).

Kendall's Coefficient of Concordance, W

Up to this point two coefficients, ρ and τ, for measuring the correlation between *two* sets of rankings of n objects or persons have been discussed. A coefficient which provides a measure of the overall correlation when there are k sets of rankings of the same n objects is Kendall's coefficient of concordance, W. Put in another way it provides a measure of the agreement between the rankings of n objects by k different judges. This coefficient is defined as:

$$W = \frac{S}{k^2(n^3 - n)/12} = \frac{12S}{k^2(n^3 - n)} \qquad \ldots 8.5$$

where S is the sum of the squares of the deviations of the total of the ranks obtained by each object from the average of these totals.

The coefficient W is the ratio of the obtained S to the maximum value S can attain. The latter would obtain were every object given the same rank by each judge and no ties were allowed. This maximum value of S can readily be shown to be $k^2(n^3 - n)/12$, for $(n^2 - 1)/12$ is the variance of the first n natural numbers, $n(n^2-n)/12 = (n^3-n)/12$ is the sum of their squares about their mean, while $k^2(n^3-n)/12$ is the sum of the squares of k times each rank about k times their mean.

The coefficient W varies from 0, signifying complete randomness in the k rankings, to $+1$, signifying complete agreement between the rankings.

Testing the significance of W. The significance of W may be tested by means of the F-distribution. To do this a correction for continuity must be applied and an adjusted value of W, say W', obtained. The adjusted value is given by the equation:

$$W' = \frac{12(S-1)}{k^2(n^3-n)+24} \qquad \ldots 8.6$$

A variance ratio, or F-value, is now obtained by the formula:

$$F = \frac{(k-1)\,W'}{1-W'} \qquad \ldots 8.7$$

and it is referred to the F-distribution with N_1 and N_2 degrees of freedom, for the horizontal and the vertical sides of the table respectively, where:

$$N_1 = (n-1)-2/k$$
and
$$N_2 = (k-1)[(n-1)-2/k] \qquad \ldots 8.8$$

In general the degrees of freedom obtained from equations 8.8 are not whole numbers so that interpolation in the F-table is necessary.

An example to illustrate W—the coefficient of concordance. The data used in this example were obtained from a 'blind' drug trial. Six patients took part in the experiment. They were administered three drugs (A) sodium amytal, (P) perphenazine, and (C) a placebo in balanced random order and for two days each. The drugs were administered in the form of tablets which were indistinguishable from each other and though the patients were told each time the treatment was changed they were not told which drug was which. Afterward the patients were asked to rank the treatments in order of preference.

The rankings are shown on the left-hand side of Table 8.3. On the right-hand side of the table the drugs are arranged in the order A, P, C, while the drug ranked first is scored 1, that ranked second is scored 2, while that ranked last is scored 3.

TABLE 8.3

Patients' Preferences for Drugs A, P and C

Patients	Rank order of drugs			Scores		
				A	P	C
I	C	P	A	3	2	1
II	P	A	C	2	1	3
III	A	P	C	1	2	3
IV	P	A	C	2	1	3
V	A	C	P	1	3	2
VI	A	P	C	1	2	3
Total				10	11	15

Grand Total = 36
Mean = 12

Deviations from mean -2 -1 3

The sum of the ranks obtained by each drug are given at the bottom of Table 8.3, together with the deviation of each sum from the average of the three. The squares of these deviations, when added, give S.

$$S = (-2)^2 + (-1)^2 + (3)^3 = 14$$

On substituting this value and the values $n = 3$, since there are three things being ranked, and $k = 6$, since there are six rankings, in equations 8.5, 8.6, 8.7 and 8.8 we obtain the values:

$$W = 0 \cdot 194$$
$$W' = 0 \cdot 176$$
$$F = 1 \cdot 068$$

while N_1 and N_2 are $1 \cdot 667$ and $8 \cdot 335$ respectively. W is very small.

When the value $1 \cdot 068$ is referred to the F-table with $1 \cdot 667$ and $8 \cdot 335$ degrees of freedom respectively it is seen not to reach any reasonable level of significance. The result, therefore, gives no reason to think that the ranking of the drugs by this sample of patients is other than random.

Incomplete Block Designs in Ranking Experiments

The unsatisfactory state of affairs which can arise when one person tries to rank a large number of people or objects has already been mentioned. The difficulty can be overcome (Durbin, 1951) by the use of incomplete block designs. These provide for the breaking down of the sample of objects or people to be ranked into sub-groups so that the members of each sub-group only are ranked at any one trial, that is on any one occasion. The sub-groups may all be ranked by the same judge or they may be allotted, preferably at random, amongst a number of judges in which case the greatest precision is obtained when each judge ranks an equal number of sub-groups.

In statistical terminology the sub-groups are referred to as blocks, and when no single block contains all the objects to be ranked it is designated an 'incomplete' block. To permit a straightforward analysis of the results obtained from the use of incomplete blocks the objects have to be allotted to the blocks in a special way. For one thing the number of times two particular objects (or people) to be ranked occur together in the same block should be the same for all pairs of objects. In addition each object should occur an equal number of times in the experiment as a whole. Moreover, should the experimenter desire to eliminate from the comparisons being made any differential effect, which, in some investigations, might arise from the order of presentation of the objects (say the order in which patients arrived for observation by the judge), this can be done by arranging things so that every object occurs equally often in each position in a block.

Designs which fulfil the first two of the above desiderata are known as *balanced* incomplete block designs. Included in these balanced incomplete blocks are some, known as Youden squares, which fulfil all three desiderata. These designs are listed by Cochran and Cox (1957), and by D. R. Cox (1957). An illustrative list from them is given below, particulars being added in an appendix at the end of the chapter. To describe the designs a few symbols are necessary. Let n be the number of objects or things to be ranked; let k be the number of objects in each block and m be the number of times each object occurs in the experiment as a whole. It follows that the number of objects (including reappearances of the same object) ranked in the experiment as a whole is mn, and as there are k objects in each block the number of blocks, b, is mn/k; in other words, $mn = bk$.

TABLE 8.4

Abbreviated List of Incomplete Block Designs including Youden Squares denoted by Y

Design No.	Number of objects to be ranked, n	Number of objects per block, k	Number of blocks, b	Number of replications m	
1	any n	2	$\frac{1}{2}n(n-1)$	$(n-1)$	
2	4	3	4	3	(Y)
3	5	3	10	6	
4	6	3	10	5	
5	7	3	7	3	(Y)
6	10	3	30	9	
7	13	4	13	4	(Y)
8	11	5	11	5	(Y)
9	21	5	21	5	(Y)

The coefficient of concordance. The formula for this coefficient when incomplete blocks are employed (and ties are absent) is given by Durbin (1951) as:

$$W = \frac{12S}{c^2 n(n^2-1)} \qquad \ldots 8.9$$

where c is the number of blocks in which a particular pair of objects occur. Since there are $\frac{1}{2}k(k-1)$ comparisons between pairs within each block the total number of comparisons is:

$$\frac{mn}{k} \times \tfrac{1}{2}k(k-1) = \tfrac{1}{2}mn(k-1)$$

it follows that:

$$\tfrac{1}{2}cn(n-1) = \tfrac{1}{2}mn(k-1)$$

so that:

$$c = m(k-1)/(n-1) \qquad \ldots 8.10$$

To test W for significance the F-ratio:

$$F = \left[\frac{c(n+1)}{k+1} - 1 \right] \frac{W}{(1-W)} \qquad \ldots 8.11$$

is calculated and is referred to the F-distribution with $2N_1$ and $2N_2$ degrees of freedom, where:

$$N_1 = \frac{mn\left[1 - \dfrac{k+1}{c(n+1)}\right]}{2\left[\dfrac{nm}{(n-1)} - \dfrac{k}{(k-1)}\right]} - \frac{k+1}{c(n+1)} \qquad \ldots 8.12$$

and

$$N_2 = \left[\frac{c(n+1)}{k+1} - 1\right]N_1 \qquad \ldots 8.13$$

An example. Part of the training of a class of students consisted in discussing at intervals in the presence of a tutor topics relevant to their work. After each discussion the tutor in charge ranked the students in order of their contribution to the discussion. Experience showed that for the purpose of these discussions small groups of students, say four at a time, were to be preferred if each student was to be given an opportunity to make his contribution. But it was also desirable that the composition of the groups should be allowed to vary in such a way that each student would, in the course of his training, find himself in a group which contained any other particular student, so that each would have a chance of joining in debate with every member of his class. Under this arrangement the tutors too would have an opportunity of ranking combinations of students which allowed all desirable comparisons to be made.

In a particular year the class consisted of 13 members. To arrange them in blocks of four the incomplete block design number (7), Table 8.4, was used. Representing the 13 students by the letters a to m this design is:

						Blocks						
1	2	3	4	5	6	7	8	9	10	11	12	13
a	b	c	d	e	f	g	h	i	j	k	l	m
b	c	d	e	f	g	h	i	j	k	l	m	a
d	e	f	g	h	i	j	k	l	m	a	b	c
j	k	l	m	a	b	c	d	e	f	g	h	i

The students were now allotted at random to the letters. The blocks too were allotted at random to the tutors—three in the present instance, two taking four debates each and one taking five. Thirteen topics for debate too were agreed upon by the tutors at the beginning of the year and these were allotted randomly to the thirteen blocks of the design.

After each discussion the tutor in charge ranked the four students concerned and at the end of the course these ranks were tabulated; they were found to be as follows:

	Students													
	a	b	c	d	e	f	g	h	i	j	k	l	m	
Ranks	1	3	1	2	2	2	3	4	3	4	3	3	2	
	3	4	1	4	1	2	4	1	2	4	1	4	3	
	1	2	2	4	1	1	3	4	2	3	1	3	1	
	3	3	2	3	1	4	4	2	4	2	2	4	1	
Sum of ranks	8	12	6	13	5	9	14	11	11	13	7	14	7	Total = 130
														Mean = 10
Deviation from mean	-2	2	-4	3	-5	-1	4	1	1	3	-3	4	-3	Total = 0
Deviation squared	4	4	16	9	25	1	16	1	1	9	9	16	9	Total = 120

From the above data it is seen that the sum of the squares of the deviations of the total rank score obtained by each student from the mean of the totals for all students, that is the quantity S, is 120. Substituting this value, together with the values $m = 4$, $n = 13$ and $k = 4$, in equations 8.9 to 8.13 we obtain:

$$W = 0 \cdot 659$$
$$c = 1 \cdot 0$$
$$F = 3 \cdot 479$$

and N_1 and N_2 equal to $5 \cdot 214$ and $9 \cdot 385$ respectively. On referring the obtained value of F to the F-distribution with $2N_1$ and $2N_2$ degrees of freedom it is found to be significant beyond the 1 per cent level. This may be interpreted as meaning that the ranking of the students by the tutors is not random, and that there is a good measure of agreement between the tutors.

If it is desired to rank the students as regards their ability as displayed in the debates, then the total rank score obtained by each is the appropriate measure to use. Under this criterion the best student is student e.

Paired Comparisons

Incomplete block designs can also be used, as Durbin (1951) points out, for paired comparisons. Indeed they are very desirable in such cases, otherwise the task of comparing objects two at a time becomes extremely tedious. When the order of presentation of the objects is immaterial Design 1, Table 8.4, is the one required. If the order of presentation of the objects is important and requires to be balanced then a Youden square must be used and the choice of designs is very limited (see Cox, 1958, p. 222). Resort must then be had to more general methods given by Kendall (1948), and Slater (1960).

To obtain the necessary symmetry for paired comparisons when Design 1, Table 8.4, is used Durbin indicates that n must be even: k, of course, is equal to two.

REFERENCES

COCHRAN, W. G. & COX, G. M. (1957) *Experimental Design* (2nd ed.), New York, Wiley & Sons

COX, D. R. (1958) *Planning of Experiments*, New York, Wiley & Sons

DURBIN, J. (1951) 'Incomplete blocks in ranking experiments', *British Journal of Statistical Psychology*, 4, 85–92

KENDALL, M. G. (1948) *Rank Correlation Methods*, London, Griffin & Co.

SCHAEFFER, M. S. & LEVITT, E. E. (1956) 'Concerning Kendall's tau, a nonparametric correlation coefficient', *Psychological Bulletin*, 53, 338–46

SLATER, P. (1960) *The Reliability of Some Methods of Multiple Comparison in Psychological Experiments*, Ph.D. Thesis, University of London

APPENDIX

Particulars of incomplete block designs given on page 123. In this appendix the people or objects to be ranked are denoted by the letters a, b, . . ., of the alphabet, and the blocks are separated by semi-colons. The number of things to be ranked is denoted by n and the number in each block by k.

Before using a design the order of the blocks should be randomised, and the randomisation should be repeated for each new replication of the experiment if replications are thought necessary for increasing the precision of the experiment. The order of letters within blocks should also be randomised except in the case of Youden squares where the order in which the letters appear in the blocks is part of the balanced properties of the design. The allocation of objects to letters or of letters, as labels, to people should also be done in a random way.

Design No.	n	k	BLOCKS
2	4	3	$a\ b\ c$; $a\ b\ d$; $a\ c\ d$; $b\ c\ d$
3	5	3	$a\ b\ c$; $a\ b\ d$; $a\ b\ e$; $a\ c\ d$; $a\ c\ e$; $a\ d\ e$; $b\ c\ d$; $b\ c\ e$; $b\ d\ e$; $c\ d\ e$
4	6	3	$a\ b\ e$; $a\ b\ f$; $a\ c\ d$; $a\ c\ f$; $a\ d\ e$; $b\ c\ d$; $b\ c\ e$; $b\ d\ f$; $c\ e\ f$; $d\ e\ f$
5	7	3	given by Durbin in his article.
6	10	3	$a\ b\ c$; $a\ b\ d$; $a\ c\ e$; $a\ d\ f$; $a\ e\ g$; $a\ f\ h$; $a\ g\ i$; $a\ h\ j$; $a\ i\ j$; $b\ c\ f$; $b\ d\ j$; $b\ e\ h$; $b\ e\ i$; $b\ f\ g$; $b\ g\ i$; $b\ h\ j$; $c\ d\ g$; $c\ d\ h$; $c\ e\ f$; $c\ g\ j$; $c\ h\ i$; $c\ i\ j$; $d\ e\ i$; $d\ e\ j$; $d\ f\ i$; $d\ g\ h$; $e\ f\ j$; $e\ g\ h$; $f\ g\ j$; $f\ h\ i$
7	13	4	used in the text, page 125.
8	11	5	$a\ b\ c\ \dots$ etc. $e\ f\ g\ \dots$ $f\ g\ h\ \dots$ $g\ h\ i\ \dots$ $i\ j\ k\ \dots$
9	21	5	$a\ b\ c\ \dots$ etc. $b\ c\ d\ \dots$ $e\ f\ g\ \dots$ $o\ p\ q\ \dots$ $q\ r\ s\ \dots$

CHAPTER IX

Miscellaneous Tests of Significance

Introduction. In this chapter a number of tests for comparing treatments or subjects' appraisal of treatments, where the responses are of the 'preferred not-preferred', 'better no-better' type, and fixed size samples are used, are described. The first is Fisher's *sign* test which is well known and widely used. Closely associated with it, as Cox (1958b, pp. 562–5) has shown, is McNemar's test (1955, pp. 228–31) for 2×2 contingency tables—already described (Chapter 1)—when the variables are related or the samples being compared are matched. When there are more than two related variables or more than two matched samples McNemar's test can be supplemented by a test given by Cochran (1950).

An interesting new series of tests too will be mentioned. They are due to Cox (1958) and are based on the logistic function. Three of these, which deal with serial data such as is often obtained from learning experiments, are of especial interest to psychologists and other behavioural scientists. Indeed they are amongst the very first tests to be constructed with the problems of human behaviour foremost in mind and it is hoped that they are the forerunners of many similar tests.

The Sign Test

This test is of wide application and though it will be illustrated here by an example in which two treatments are compared the reader should easily appreciate the great variety of investigations, of similar design, in which it can usefully be employed. The sign test is one for use with qualitative data. It can be applied to quantitative data but it is less powerful in such cases than the *t*-test for correlated means—a sample size of 63 when the t-test is employed being equivalent to one of 100 when the sign test is used.

In an experiment a placebo and a tranquillising drug were administered, in tablets of similar appearance, in random order to each of 27 patients. Though the patients knew that two different treatments were being administered they did not know which was being administered in the first and which in the second part of the experiment. Each treatment lasted three days and at the end of the experiment the patients had to say which of the two had done them the more good. (Another way of doing the experiment would have been to match the patients in pairs before the experiment and then allot one of each pair at random to each treatment. In this case the judgment as to which treatment was preferable could be made by a third person, who like the patients did not know which treatment was which.)

Of the 27 patients two were unable to decide between the treatments. As they provided no information about the relative merits of each they were dropped from the experiment. For the remaining 25 patients a '+' sign was recorded if the drug was found to be preferred and a '−' sign if the placebo was preferred. The experiment yielded 18 pluses and 7 minuses. The statistical problem was to decide whether this result is indicative of a real difference between the treatments.

The sign test is based on the binomial theorem. On the null hypothesis that the treatments are of equal benefit one would expect the experiment to yield pluses and minuses in roughly equal numbers. The binomial theorem allows us to calculate the probability of any given number of signs of one kind or the other occurring. This information is obtained from the terms in the expansion of the binomial expression $(\frac{1}{2} + \frac{1}{2})^n$ where n is the number of subjects (or the number of pairs of subjects) in the experiment, i.e. is the number of comparisons made. The expansion is:

$$(\tfrac{1}{2} + \tfrac{1}{2})^n = \tfrac{1}{2}^n[1 + n + n(n-1)/2! + n(n-1)(n-2)/3! + \ldots$$
$$+ n(n-1)/2! + n + 1] \qquad \ldots 9.1$$

The expansion has $(n+1)$ terms and is symmetrical. The first term gives the probability that all n signs are alike of one kind, while the last term—which is equal to it—gives the probability that all terms are alike and of the other kind. The second term gives the probability that all the signs save one are alike of one kind, while the second last term—which is equal to it—gives the probability that all the signs save one are alike and of the other kind; and so on for the other terms equidistant from the beginning and the end of the series.

The probability of getting k or less signs alike of one kind and $(n-k)$ or more alike of the other kind is given by the sum of the first (or last) $k+1$ terms of the series. Here the question of one-tail or two-tail tests of significance enters. If the hypothesis to be tested is simply that the treatments differ (a two-tail test) then the sum of both the first and the last $(k+1)$ terms of the series—or what is the same thing, twice the sum of the first $(k+1)$ terms—is required. On the other hand, if the hypothesis being tested is that the drug is more effective than the placebo (one-tail test) then we require the sum of the first $(k+1)$ terms only.

In our example we are concerned with the two-tailed hypothesis. Since n is 25 and k is 7 the probability required is given by twice the sum of the first 8 terms of the series, namely:

$$2[\tfrac{1}{2}^{25}(1+25+(25\times24)/2!+(25\times24\times23)/3!+ \text{ to 8 terms]}$$

$$= \frac{1}{16777216}(1+25+300+2300+\ldots = 0\cdot0433$$

Since this probability is just less than the 5 per cent level ($P < 0\cdot05$) the null hypothesis may be rejected and the result taken as evidence that the drug is preferable to the placebo.

Easing the calculations. The calculations in the example given are onerous. This is generally so when n is fairly large and k is greater than 3 or 4. However, it often happens that the sum of the first few terms of the series exceeds the value $P = 0\cdot05$, so that the null hypothesis can be accepted without calculating the remaining terms. Moreover, when n is about 20 or more a good approximation to the true probability is given by a test based on the normal approximation to the binomial theorem. To do it the expression:

$$z = \frac{(|k-n/2|-0\cdot5)}{0\cdot5\sqrt{n}} \qquad \ldots 9.2$$

is evaluated. The vertical lines in the numerator tell us to take the value of $(k - n/2)$ as positive whether or not it is so algebraically. The sample size, or rather the number of subjects in the experiment who give a definite positive or negative answer, is denoted by n, and k is the number of pluses or the number of minuses, whichever is the smaller. The term $0 \cdot 5$ in the numerator of expression $9 \cdot 2$ is a continuity correction analogous to the Yates correction used in chi-square tests.

For the example above $n = 25$, and $k = 7$, so that:

$$z = \frac{|7 - 12 \cdot 5| - 0 \cdot 5}{0 \cdot 5 \times \sqrt{25}} = \frac{5}{2 \cdot 5} = 2 \cdot 0$$

This value, when referred to the normal curve, corresponds to a probability of $P = 0 \cdot 0456$ and so agrees well with the earlier result where the true value of P was found to be $0 \cdot 0433$.

Cochran's Q-Test for Correlated Quantal Results

The test just considered, together with McNemar's test, Chapter 1, for correlated data in a fourfold table, have their natural extension in a test by Cochran (1950) which enables one to compare k qualitative ratings for each of a number of subjects, or one rating for each of a number of sub-groups of k matched subjects. To clarify the matter suppose that patients are tested for three consecutive weeks on three different drugs. For the first two days of each week a drug is given but for the remainder of the week it is withheld. At the end of the week each patient has to say whether or not he felt better when having the drug than when without it. When a preference is stated for the drug a '1' is scored, otherwise a '0' is scored. Since there are six possible orders in which three treatments A, B, C, (one of which incidentally is generally a placebo included for purposes of control) can be given, namely ABC, BCA, CAB, CBA, ACB, and BAC, the patients should be taken in groups of six and allotted at random to the six different orders. Suppose that for two replications of the experiment, that is for two sets of six patients, the results in Table 9.1 are obtained. The problem then is to test the null hypothesis that the proportion of preferences

TABLE 9.1

Preferences for Drug or No-Drug

Patient	Drug A	Drug B	Drug C	Total	(Total)²
1	1	0	1	2	4
2	1	1	0	2	4
3	1	1	0	2	4
4	1	1	1	3	9
5	1	0	1	2	4
6	0	0	0	0	0
7	1	0	1	2	4
8	1	1	1	3	9
9	0	1	0	1	1
10	1	1	0	2	4
11	1	0	1	2	4
12	1	1	1	3	9
Total	10	7	7	24	56

for a drug as opposed to no drug is the same whichever drug is considered. To make the test Cochran has shown that the expression:

$$\frac{k(k-1)\sum c_j^2}{k\sum r_i - \sum r_i^2} \qquad \ldots 9.3$$

is distributed approximately as chi square with $(k-1)$ degrees of freedom. Referring to Table 9.1, k in the formula is the number of treatments—drugs in our case. r_i is the sum of the scores for the i-th patient, while c_j is the deviation of the sum of the scores for the j-th treatment from the mean of the sum of all k treatments.

For the data in Table 9.1 $\sum r_i = 24$, $\sum r_i^2 = 56$, while

$$\sum c_j^2 = 10^2 + 7^2 + 7^2 - 24^2/3 = 6$$

hence $\chi^2 = 2\cdot25$. With two degrees of freedom this value is not significant, so that there are no grounds for claiming that any one drug is preferred to another. In conclusion we may note that since the responses given by patients 4, 6, 8 and 12 do not contribute towards differentiating between the drugs these patients might have been omitted when analysing the results. This fact enables us to make a check on the test and if the student reworks the data omitting these four patients he will find that the value of χ^2 obtained is $2\cdot25$, as before.

Tests for Serial Data

The next three tests to be described (Cox, 1958a) are of especial value for analysing serial data such as are frequently obtained in learning and similar experiments. The first is a *test for trend*.

Suppose that an experimental task is attempted by a subject n times in succession and that of two possible responses at each trial the correct one is determined in advance by the experimenter by randomisation. The task might be one of the single maze type in which at each choice-point the correct turn, left or right, had previously been decided upon, when the maze was being constructed, by the toss of a coin. Or it might be one of the kind often employed in the information theory type of experiment in which the subject has to predict on each of a number of successive trials whether or not a symbol, say a light, will appear after a given signal, when the probability that the symbol will appear has previously been fixed by the experimenter.

Suppose that in an experiment of the former type, in which the probability of a left or a right turn being the correct response is $\frac{1}{2}$, the following series—in which S stands for a 'success' and F for a 'failure' — is obtained:

Trial number	1	2	3	4	5	6	7	8	9	10	11	12	13	14	15
Response	F	F	F	S	F	S	S	F	F	S	S	S	F	S	F

Trial number	16	17	18	19	20	21	22	23	24	25	26	27	28	29	30
Response	S	S	S	S	F	S	S	F	S	S	F	S	F	F	S

Then we might wish to test the null hypothesis that successes occur randomly with probability $\frac{1}{2}$ against the alternative hypothesis that there is a trend in the data. This is a test of regression on serial order. To do it the trials are numbered from 1 to n—from 1 to 30 on our example—and the sum of the trial numbers corresponding to 'successes' in the series is obtained. For the data in question, this sum, which will be denoted by Y, is:

$$Y = 4 + 6 + 7 + 10 + 11 + 12 + 14 + 16 + 17 + 18 + 19 + 21 +$$
$$+ 22 + 24 + 25 + 27 + 30$$
$$= 283$$

Cox has shown that the theoretical mean M of Y, that is the expected value of Y, on the null hypothesis when the probability of a success in any trial is p, is given by:

$$M = pn(n+1)/2 \qquad \qquad \ldots 9.4$$

while the theoretical variance, V, is given by:

$$V = np(1-p)(n+1)(2n+1)/6 \qquad \ldots 9.5$$

To do the test the ratio:

$$z = (Y-M)/\sqrt{V} \qquad \qquad \ldots 9.6$$

is referred to the normal curve. However, when n is small a correction for continuity should be applied if the assumption that z is normally distributed is made. The corrected value, z', of z is given by:

$$z' = (|Y-M| - 0 \cdot 5)\sqrt{V} \qquad \ldots 9.6a$$

For our example n is 30 and p is $\frac{1}{2}$. On substituting these values in equations 9.5 and 9.6 we obtain:

$$M = \tfrac{1}{4} \times 30 \times 31 = 232 \cdot 5$$

and

$$V = (30 \times 31 \times 61)/(4 \times 6) = 2363 \cdot 75$$

hence

$$z' = (|283 - 232 \cdot 5| - 0 \cdot 5)/\sqrt{(2363 \cdot 75)}$$
$$= 1 \cdot 028$$

This value, when referred to the normal curve, is found to correspond to a probability of $P = 0 \cdot 304$ (two-tail) which would not lead us to reject the null hypothesis that successes occurred randomly with probability equal to one-half.

Cox's Cumulative Score Test

Here the problem is one of testing the hypothesis that the probability of a success, S, on the i-th trial in a learning task is a function of the number of successes achieved on the previous $(i-1)$ trials—but, as will be understood, the test will be found to be applicable in situations other than that envisaged here. To illustrate the procedure let us take the results of the first 20 of a long series of trials from a learning experiment in which successes are scored '1' and failures are scored '0'.

Number of trial	1	2	3	4	5	6	7	8	9	10
Response	0	0	0	1	0	1	1	0	0	1

Number of trial	11	12	13	14	15	16	17	18	19	20
Response	1	1	0	1	0	1	1	1	1	0

Let the number of successes be denoted by y; then $y = 11$. Now let the serial number of the first success be denoted by r_0, the serial number of the second success by $r_0 + r_1$, of the third by $r_0 + r_1 + r_2$, and so on. Equating these combinations of r's to the serial numbers in question a series of equations is obtained from which the values of r_0, r_1, ... $r_{(y-1)}$, can be evaluated. For example for the series of trials given above:

$$r_0 = 4$$

since the first success is on the fourth trial. Similarly $r_0 + r_1 = 6$, since the second success is on the sixth trial. In the same way we find that:

$$r_0 + r_1 + r_2 = 7$$
$$r_0 + r_1 + r_2 + r_3 = 10$$

and so on to $\quad r_0 + r_1 + \dots + r_{10} = 19$

for the 11-th or last success. From these equations we know that $r_0 = 4$, hence $r_1 = 2$, so that $r_2 = 1$, etc. We then have r_1, r_6 and r_7 equal to 2; r_3 equal to 3; and r_2, r_4, r_5, r_8, r_9 and r_{10} each equal to 1.

One further r-value, namely r_k—where k is the number of successes—is now obtained: this brings the number of r's to $(k+1)$ since the series began with r_0. It is given by the equation:

$$r_k = n - r_0 - r_1 - r_2 - \dots - r_{(k-1)} \qquad \dots 9.7$$

where n is the number of trials. For the present example r_k is:

$$r_{11} = 20 - 4 - 2 - 1 - 3 - \dots - 1 = 1$$

In cases where the sum of the k terms r_0 to $r_{(k-1)}$ is equal to n then r_k will be zero, otherwise it will be a positive integer, as it is in the present example.

The series of $(k+1)$ r's is now tabulated together with their arithmetic values calculated above. They are:

Serial r	r_0	r_1	r_2	r_3	r_4	r_5	r_6	r_7	r_8	r_9	r_{10}	r_{11}
r-value	4	2	1	3	1	1	2	2	1	1	1	1

The test statistic employed for testing the hypothesis that the probability of a success on the i-th trial is a function of the number of successes achieved on the previous $(i-1)$ trials is:

$$R = \sum k r_k \qquad \qquad \ldots 9.8$$

that is the sum of products of the r-values and their serial numbers, i.e. their subscript values. For the data in question:

$$R = (4 \times 0) + (2 \times 1) + (1 \times 2) + \ldots + (11 \times 1) = 86$$

The mean, M, or the expected value of R as distinct from its observed value (86 in our example) is shown by Cox to have the value:

$$M = \tfrac{1}{2}(y'-1)n \qquad \qquad \ldots 9.9$$

where $y' = y+1$ when the final value of r (equation 9.7) is non-zero, or $y' = y$, when the final r-value is zero. Recalling that for our series the number of successes is 11 (i.e. $y = 11$) and that the final r-value is non-zero we require $y' = (y+1) = 11+1 = 12$, so that the expected value of R is (equation 9.9):

$$M = \tfrac{1}{2}(y'-1)n = \tfrac{1}{2}(11 \times 20) = 110$$

The estimate of the variance of M, the expected value of R, is given by the expression:

$$V(M) = \frac{y'(y'+1)}{12} \sum (r_i - \bar{r})^2 \qquad \qquad \ldots 9.10$$

where $\bar{r}$ is the mean of the r-values. The expression $\sum (r_i - \bar{r})^2$ can most easily be evaluated by subtracting from the sum of the squares of the r's the square of their sum divided by the number of r's. For our example:

$$\sum (r_i - \bar{r})^2 = 4^2 + 2^2 + 1^2 + \ldots + 1^2 - 20^2/12$$
$$= 10 \cdot 67$$

so that:

$$V(M) = \frac{12 \times 13}{12}(10 \cdot 67) = 138 \cdot 71$$

The estimate of the error variance of M is therefore $\sqrt{(138 \cdot 71)} = 11 \cdot 778$.

To test the hypothesis that the probability of a success on any trial is a function of the number of previous successes the ratio:

$$z = \frac{|R-M| - 0 \cdot 5}{\sqrt{[V(M)]}}$$

which is approximately normally distributed, is obtained. For the data in the example:

$$z = \frac{|86 - 110| - 0 \cdot 5}{11 \cdot 778} = 2 \cdot 00$$

the vertical lines in the numerator indicating that the absolute value of $(R-M)$, that is the value irrespective of sign, is to be taken. When referred to the normal curve the value $2 \cdot 00$ is found to correspond to a probability of $P = 0 \cdot 046$. This would lead us to reject the null hypothesis so that on the results from this experiment there is some basis for claiming that success on any trial on the task in question is a function of the number of previous successes on that task.

Should we desire to combine results from p independent series then for each series R, M and $V(M)$ are evaluated. The difference $d = (R-M)$ for each series is obtained and the ratio:

$$z = \frac{\sum d_i}{\sqrt{[\sum V(M_i)]}}, \quad i = 1, 2, \ldots p$$

is referred to the normal curve.

Test of a Simple Markov Chain

This test is applicable when the problem is one of deciding whether the probability of a success in the i-th of a series of trials in an experiment, such as a learning experiment, depends on the result of the $(i-1)$th trial; hence the title. The data used in the previous section will serve for illustrating the procedure. If n is taken to represent the number of trials then $n = 20$, and y—the number of successes—is 11. To do the test the number of runs of successes in the series is also required. By a 'run' of successes is meant a group of adjacent *ones*, but a group may contain just a single *one*. For the series under consideration $W = 5$.

A fourfold table having the following entries is now constructed and a chi-square test with correction for continuity, or an exact probability test—should the data call for it—is then carried out:

$$W \qquad\qquad (y - W)$$
$$(n - y - W + 1) \qquad (W - 1)$$

Rejection of the null hypothesis would mean that success in any trial of the series was not on the average independent of the outcome at the immediately preceding trial. The student who wishes to understand the theoretical basis for the test is exhorted to refer to Cox's paper (1958a) and to a complementary paper by Stevens (1939).

REFERENCES

COX, D. R. (1958a) 'The regression analysis of binary sequences', *Journal of Royal Statistical Society* **B20**, 215–32

COX, D. R. (1958b) 'Two further applications of a model for binary regression', *Biometrika*, **45**, 562–5

COCHRAN, W. G. (1950) 'The comparison of percentages in matched samples', *Biometrika*, **37**, 256–266

MCNEMAR, Q. (1955) *Psychological Statistics*, New York, Wiley & Sons

STUART, A. (1957) 'The comparison of frequencies in matched samples', *British Journal of Statistical Psychology*, **10**, 29–32

STEVENS, W. L. (1939) 'Distribution of groups in a sequence of alternatives', *Annals of Eugenics*, **9**, 10–17

CHAPTER X

Classification Procedures based on Bayes's Theorem and Decision Theory

Introduction. Questions of identification and classification occur frequently in the biological and behavioural sciences, and in psychiatry they are of current interest since some dissatisfaction exists about diagnostic categories in present use. From the statistical viewpoint the classification problem is one of allotting, in some optimal way, members of a composite population to categories within that population. For instance the composite population may consist of people who are mentally ill and the categories may be subdivisions of this population which the psychiatrist finds useful when considering treatment and prognosis. The statistician then tries to set up decision procedures by which members drawn randomly from the composite population can 'best' be allotted to the categories on the basis of measurements made on them.

Modern text-books on multivariate analysis (Rao, 1952, for example) give several statistical procedures for dealing with classification problems when the measurements, or variables, considered are continuous. But a simple method for use when the variables are of the dichotomous type—say, the presence or absence of certain symptoms where psychiatric patients are concerned—should be helpful. In this chapter one such method is described. It has the special attraction that it allows the relative seriousness of the different types of error which can occur in any particular situation to be taken into account. This is an advantage, for current methods of classification tend tacitly to assume that all types of error are of equal importance. But clearly it could be more serious were a psychiatrist to misclassify a patient with a brain tumour as a schizophrenic than the reverse, and the statistical procedures recommended for use in classification problems should aim at taking such aspects of the problem into account. In attempting to do this the method to be described below makes use of some ideas from *statistical decision theory* (Wald, 1950; Barnard, 1953). The relevance of this theory for problems of classification has recently been considered by Birnbaum and Maxwell (1961) and the methods outlined in their paper are followed here.

The problem. The following classification problem, which is typical of the type of problem met in the biological sciences, will be used as a basis for the discussion. The composite population in this instance is composed of patients from each of the three diagnostic categories Affective Disorders, Anxiety States and Neurotic Depressives, and random samples from each of these categories—as indicated below— are drawn.

	Category	Sample size
I	Affective disorders	148
II	Anxiety states	100
III	Neurotic depressives	132
	Total	380

For these patients, all of whom incidentally are males between the ages of 16 and 59 years inclusive, the presence or absence of the following three symptoms were noted:

(a) depressed;
(b) lacks confidence when in society;
(c) showed mood variations previous to present illness.

Now if the presence of a symptom is denoted by '1' and its absence by '0' and three symptoms are considered there are 2^3, or 8, possible answer patterns or score configurations which it is possible for the patients to show. These are indicated on the left-hand side of Table 10.1; each answer pattern is thus represented by a number in the dyadic or binary number system.

Had there been n symptoms then there would have been 2^n distinct answer patterns. Inspection of the left-hand side of Table 10.1 indicates how these patterns could be written down quickly. For example the first column of the table has half its entries *zeros* and half *ones*. The second column has the first quarter of its entries *zeros* and the second quarter *ones*, the third quarter *zeros* and the fourth quarter *ones*. The third column begins with *zeros* in the first eighth of its entries, *ones* in the second eighth, and so on, for n columns.

TABLE 10.1

Answer Patterns for Three Symptoms, a, b and c and the Frequency of Occurrence of these Patterns for the Three Categories: I, Affective Disorders; II, Anxiety States; III, Neurotic Depressives

Answer pattern	Symptoms			Categories			Allocation of patterns to categories	Total
	a	b	c	I	II	III		
1	0	0	0	11	19	3	II	33
2	0	0	1	13	9	6	I	28
3	0	1	0	3	13	0	II	16
4	0	1	1	4	12	1	II	17
5	1	0	0	30	14	44	III	88
6	1	0	1	38	11	23	I	72
7	1	1	0	18	9	23	III	50
8	1	1	1	31	13	32	III	76
Total				148	100	132		380

In the centre section of Table 10.1 the frequencies with which the patterns occur in the several categories are given. These can be obtained very quickly, even for large n, if the data are punched on cards which can be sorted mechanically. Indeed the task is often lighter than it appears, for all the 2^n possible patterns will not necessarily occur for any given set of categories.

Allotting the answer patterns to the categories. Having tabulated the data as in Table 10.1 the classification procedure to be outlined is one of deciding how to allot the answer patterns to the several categories, each pattern to just one category, a so as to minimise the number of misclassifications, and b so as to keep errors, of the kinds thought to be most serious, small. Problem a will be discussed first

When the relative probabilities—the so-called prior probabilities—that a patient, randomly chosen from the composite population, belongs to one or other of the given categories are known the answer patterns can be allotted to the categories by resort to Bayes's theorem. In the case of the categories comprising our hypothetical composite population records for the hospital concerned show these prior probabilities to be:

	Category	Prior probability (g_i)
I	Affective disorders	0·40
II	Anxiety states	0·25
III	Neurotic depressives	0·35
	Total	1·00

In other situations, of course, it might not be possible to ascertain the required prior probabilities but, as will be seen when we come to discuss problem *b*, this fact would not preclude the use of the procedures outlined.

Knowing the prior (often referred to as the *a priori*) probabilities the next stage in the analysis is to express the frequencies in Table 10.1 as proportions, though this would not be necessary were the sample sizes equal. These proportions are shown in columns I, II and III respectively of Table 10.2. Now if *u* is chosen as a generic term to represent answer patterns, and *i* to represent categories, these proportions can be referred to symbolically as $p(u;i)$—the probability that a patient drawn randomly from category *i* will have the answer pattern *u*. By the use of Bayes's formula we can now calculate *a posteriori* probabilities, $p(i;u)$, three for each answer pattern in our case—that is, one for each category. These posterior probabilities are the probabilities that a patient drawn randomly from the composite population and found to have the answer pattern *u* comes from Category *i*. Bayes's formula is:

$$p(i;u) = \frac{g_i.p(u;i)}{\sum g_i.p(u;i)} \qquad \ldots 10.1$$

where g_i is the prior probability for the *i*-th category.

Luckily the denominator in this formula need not be calculated as it is the same for each category for a given answer pattern. Hence we can take $p(i;u)$ as being proportional to $g_i . p(u;i)$. The latter values are given in Table 10.2 for the present example. The entries in the fourth column of that table, for example, are got by multiplying the entries in the first column by $g_I = 0 \cdot 40$; those in column five by multiplying those in column two by $g_{II} = 0 \cdot 25$, etc.

TABLE 10.2

The Probabilities $p(u;i)$ and the Weighted Probabilities $g_i \times p(u;i)$
for the Categories I to III inclusive

Pattern No.	p(u;i)			$g_i \times p(u;i)$			Pattern allocated to category number
	I	II	III	I	II	III	
1	0·074	0·190	0·023	0·030	0·048	0·008	II
2	0·088	0·090	0·045	0·035	0·022	0·016	I
3	0·020	0·130	0·000	0·008	0·032	0·000	II
4	0·027	0·120	0·008	0·011	0 030	0·003	II
5	0·203	0·140	0·333	0·081	0·035	0·116	III
6	0·257	0·110	0·174	0·103	0·028	0·060	I
7	0·122	0·090	0·174	0·049	0·022	0·061	III
8	0·209	0·130	0·242	0·084	0·032	0·085	III

The entries in each row of the weighted probabilities in Table 10.2 are now inspected and the answer pattern concerned is allotted to the category for which the weighted probability is largest. The allocations of the answer patterns are shown in the final column of the table. *A consequence of the use of Bayes's theorem when allocating the patterns is that it ensures that the number of misclassifications is a minimum.*

Appraising the Results

A table showing the number of correct (principal diagonal) and incorrect classifications is now drawn up—Table 10.3. The first entry in the table, viz. 51, is the sum of the frequencies in column four of Table 10.1 for the patterns allocated to Category I, namely (13 + 38). The second entry of 18 in the first row of Table 10.3 is the sum of the frequencies in column four of Table 10.1 for answer patterns allotted to category II, namely (11 + 3 + 4). In a similar manner the other entries in Table 10.3 are found. A check on the calculations is that the sums of the rows of the latter table must equal the sample sizes for the respective categories.

If the entries in the leading diagonal of Table 10.3(i) are added we obtain 194 which, when expressed as a percentage of 380, the total sample size, gives 51 per cent. This is the percentage of patients correctly classified. The advantage derived from using the information supplied by the presence or absence of the three symptoms considered can now be assessed. Looking again at the *a priori* probabilities for the categories, namely 0·40 for Affective Disorders, 0·25 for Anxiety States and 0·35 for Neurotic Depressives it is clear that if

TABLE 10.3

(i) *Numbers of Patients Correctly and Incorrectly Classified, and* (ii) *Conditional Probabilities of Correct and Incorrect Classification using the Initial a priori Probabilities*

	(i)				(ii)			
	I	II	III	Total	I	II	III	Total
Categories I	(51)	18	79	148	0·34	0·12	0·53	1·00
II	20	(44)	36	100	0·20	0·44	0·36	1·00
III	29	4	(99)	132	0·22	0·03	0·75	1·00

patients had to be allotted to one or other of these categories in the absence of all other information about them save the knowledge given by these *a priori* probabilities the best procedure to adopt would be to allocate every patient to the first category. In this way 40 per cent of the patients would be correctly classified. But utilising the knowledge given by the three symptoms, in addition to that provided by the *a priori* probabilities, 51 per cent of the patients were correctly classified. The improvement due to the knowledge afforded by the symptoms is then 11 per cent more patients correctly classified. Further improvement could be achieved by increasing the number of symptoms in the study, but of this more later. For the moment it will be well to look again at Table 10.3(i). If the entries in each row are divided by the row total the proportions, or conditional probabilities, given in Table 10.3(ii) are obtained. The first entry, 0·34, is the probability that given a patient is from Category I he will be correctly assigned to Category I. Similarly for the other principal diagonal entries in the table. The off-diagonal entries are the conditional probabilities of misclassification. For instance, the probability that a patient from Category III will be misclassified as belonging to Category I is $0 \cdot 22$.

It is now advisable to inspect Table 10.3 more carefully to see where most misclassifications occur, and to consider the advisability—in the light of the relative seriousness of the different types of errors involved—of adjusting the decision procedure adopted. Such considerations lead to the second problem outlined earlier in this chapter, and incidentally to the procedure to be adopted in cases where *a priori* probabilities are not available. It is in the consideration of these problems that *statistical decision theory* is especially helpful.

Statistical Decision Theory

While a full discussion of statistical decision theory is far beyond the scope of this book, its relevance for classification problems is relatively easy to appraise and has been discussed in a non-technical way by Birnbaum and Maxwell (1959). Briefly the position is this. The rules by which the answer patterns, u, are allocated to the diagnostic categories, i, are referred to as decision rules, or decision functions, and are generally denoted symbolically by $d(u)$. A decision function such as that suggested by Bayes's theorem, employed above, which indicated that each answer pattern should be allocated to the category for which its *a posteriori* probability, $p(i; u)$, was greatest is called an *admissible* decision function—admissible because no other rule could have given fewer misclassifications in the circumstances. For this reason any other rule which might have been entertained for allotting the answer patterns would have been less efficient and therefore *inadmissible*. Now in statistical decision theory the general procedure for arriving at admissible decision functions is by way of loss functions. Assuming that the relative seriousness of the possible errors of misclassification could be measured along some common scale of utility, a loss function is set up and the decision functions eventually adopted are chosen so as to minimise in some acceptable way the losses involved.

In many common decision problems it is quite impractical to try to set up an appropriate loss function on account of the number and variety of imponderables which affect the situation. Unfortunately— as is pointed out elsewhere (Birnbaum and Maxwell, 1959)—'much writing in the field of statistical decision theory fails to mention the fact that even when such loss functions are quite hypothetical from the standpoint of the situation of application the decision functions determined by the formal use of hypothetically and arbitrarily-defined loss functions will typically have the property of admissibility when they are appraised more directly and simply in terms of the probabilities of the errors of the various sorts to which their use leads. A notable exception is the paper by Lindley (1953).'

Constructing admissible decision functions. The basic general method of constructing the various admissible decision functions is to introduce, in a hypothetical formal way, the various possible fictitious *a priori* probabilities that an individual belongs to Category *i*. To keep the discussion concrete let us return to our example. On examination of the conditional probabilities in Table 10.3(i) it is seen that, given a patient comes from Category I, Affective Disorders, the probability that he will be classified—by the decision rules employed when obtaining the results given in that table—as a Neurotic Depressive, Category III, is 0·53. This probability is very high, considerably higher than the probability (0·43) that the patient in question will be correctly classified. It would not be surprising therefore if a psychiatrist examining our classification procedure were to express dissatisfaction with it. He might point out that the misclassification of an Affective Disorder is likely to be more serious than of a Neurotic Depressive, yet our scheme correctly classifies 75 per cent of the latter and only 34 per cent of the former. The statistician would now have some indication of the relative seriousness, or cost, of the misclassifications concerned. To utilise this information when allotting 'answer patterns' he would need to augment equation 10.1. This could be done by a system of weighting chosen to reduce the number of misclassifications of the more serious type. At the arithmetic level it so happens that the introduction of such weights can be effected by appropriately altering the numerical values of the quantities representing the prior probabilities, to give new 'hypothetical' prior probabilities.

CLASSIFICATION PROCEDURES 147

For our data the prior probabilities are 0·40, 0·25 and 0·35 for
Affective Disorders, Anxiety States and Neurotic Depressives respectively. In order to classify more of the Affective Disorders correctly the numerical value of the prior probability for this category
would require to be bigger than it is. But as the sum of the prior
probabilities must be unity an increase in one will necessitate a decrease in one or more of the others. In view of the high percentage of
correct classifications which our original strategy achieved with
Neurotic Depressives we may consider lowering the value of the prior
probability for this category. Suppose then that the adjusted values
of the g's be taken as 0·46, 0·25 and 0·39 (which add to unity) for

TABLE 10.4

(i) *Numbers of Patients Correctly and Incorrectly Classified, and*
(ii) *Conditional Probabilities of Correct and Incorrect Classification
using Adjusted Prior Probabilities*

		(i)				(ii)		
	I	II	III	Total	I	II	III	Total
Categories I	(82)	18	48	148	0·55	0·12	0·32	1·00
II	33	(44)	23	100	0·33	0·44	0·23	1·00
III	61	4	(67)	132	0·46	0·03	0·50	1·00

(Principal diagonal entries shown in brackets)

the three categories respectively. The entries in Table 10.2 can now
be recalculated and the 'answer patterns' reallotted amongst the
groups, using Bayes's formula. The reallocation of the answer
patterns in turn from 1 to 8 now becomes II, I, II, II, III, I, III, I.
Data equivalent to that given in Table 10.3 can now be obtained for
the new allocation. These are given in Table 10.4.

The percentage of correct classifications has not changed noticeably; to the nearest whole number it is 51 per cent, as before. But the
conditional probability that an Affective Disorder will be classified
as such has now increased from 0·34 to 0·55, while the probability
that he will be misclassified as a Neurotic Depressive has fallen from
0·53 to 0·32. This improvement, as will be seen by comparing the
conditional probabilities in Tables 10.3 and 10.4, is achieved at a
price. To mention only one case the probability that a Neurotic

Depressive will be misclassified as an Affective Disorder has increased from $0 \cdot 22$ to $0 \cdot 46$. The new results, however, are more in keeping with what we have assumed to be the psychiatrist's notions about the relative seriousness of the different types of errors which can occur when the proposed procedure is used. The adjusted *a priori* probabilities may therefore be considered preferable to the real ones in the sense that they lead to more satisfactory decision rules.

From the above discussion it follows that prior probabilities need not be viewed as sacrosanct when Bayes's theorem is viewed in the light of statistical decision theory. Indeed they are not even necessary. In their absence equal probabilities for the categories could first be assumed and these could then be suitably adjusted after an examination of the errors of misclassification to which their use led.

Apart too from the question of the ultimate choice of prior probabilities the procedures outlined in this chapter have a research aspect. This concerns the actual choice of symptoms, or items, to use in the first place. Here there is no general rule to guide us so it would be necessary to try out various symptoms and combinations of symptoms until an efficient set was found.

Advantages and disadvantages of the procedure. Finally, it is desirable that the advantages and disadvantages of the classification method based on Bayes's formula and outlined in the preceding paragraphs should be clearly stated. The principal advantage of the method is that no restrictive theoretical assumptions are made about the distribution of the probabilities $p(u;i)$ estimated from the samples. All that is required is that the samples be randomly drawn. Another advantage of the method is that the calculations involved are very simple and straightforward.

The main disadvantage is that as the number of variables (symptoms, items, etc.) is increased the number of possible answer patterns, 2^n, increases exponentially. This means that for even a moderate number of variables very large samples of subjects would be required to determine the probabilities $p(u;i)$ with reasonable accuracy. However, this difficulty could be overcome in part by considering only a few variables at a time and in this way singling out those which were the most efficient for discriminative purposes.

In conclusion it is well to note that the results given by an application of the method to a single set of random samples, say those given in Tables 10.3 and 10.4, almost certainly give too glowing an impression of the efficiency of the method. Its true efficiency would have to be assessed by trying out the decision rules derived from one study on further random samples from the population and recalculating the conditional probabilities of correct and incorrect classifications which resulted.

REFERENCES

BARNARD, G. A. (1953) Review of A. Wald's *Statistical Decision Functions* in *Biometrika*, **40**, 475–7

BIRNBAUM, A. & MAXWELL, A. E. (1961) 'Classification procedures based on Bayes's Formula', *Applied Statistics*, 9

LINDLEY, D. V. (1953) 'Statistical inference', *Journal of the Royal Statistical Society*, **B15**, 30–76

RAO, C. R. (1952) *Advanced Statistical Methods*, New York, Wiley & Sons

WALD, A. (1950) *Statistical Decision Functions*, New York, Wiley & Sons

CHAPTER XI

Item Analysis and the Construction of Attitude Scales

Introduction. A knowledge of people's beliefs and attitudes is considered today to be essential to successful industrial and commercial enterprise. As a result great interest is now shown by market research firms and advertising and information services in techniques for measuring attitudes, interests, etc. The notion that such things can be measured is, therefore, no longer new to the man in the street. For psychologists the problem is an old one and a statistician working with psychiatrists, psychologists or sociologists is regularly approached for advice on how to tackle some problem of attitude measurement—the attitude of children to their parents, of patients to therapists, of psychiatric colleagues to Freudian analysis or to new drugs, and so on.

Questions of attitude measurement are difficult. The main problems are these—attitudes, like most other psychological phenomena can only be measured indirectly; they are difficult to define and vary greatly from culture to culture and often within different cultures themselves and from one socio-economic group to another. The problem is further complicated by what may be termed the 'uncertainty principle' in psychology—the irritating fact that any approach to a person aimed at probing him about his attitudes may engender in him a reaction, often covert, which renders an unbiased assessment of his true attitude difficult or even impossible. It is possibly in an effort to overcome such barriers that there has grown up a notion that the measurement of an attitude requires a scale or nucleus of questions, rather than just one or two direct questions coming straight to the point, for its measurement. By a whole series of questions carefully chosen and not too direct one hopes to circumvent the natural reticence, if not the involuntary inhibition, of the subject and obtain from him information of the kind required.

Attitude inventories. In the preface to a valuable 250-page book recently published and devoted exclusively to a description of techniques of attitude scale construction the author (Edwards, 1957) is at pains to point out that his account of the subject is by no means exhaustive; so obviously any treatment of the subject which is confined to just one short chapter necessitates stringent selection of the topics of attitude measurement to be treated. As a result attention here is confined to the statistical questions which arise in the construction of the common attitude questionnaire or personality inventory. It generally consists of a number of questions or items, as they shall be called, of a homogeneous type, and has its counterpart in the field of cognitive tests in the familiar group test of the Moray House type. As a consequence the very considerable body of statistical knowledge which has grown up over the years concerning the construction of such tests (Ferguson, 1942; Lawley, 1942; Finney, 1944; Lord, 1953; Birnbaum, 1957; Maxwell, 1959) is at our disposal when we come to construct attitude inventories, and reference will be made to it shortly. Some preliminary questions, however, should be dealt with first.

It is outside the scope of this book to try to define an attitude and the word will be taken to have its ordinary everyday meaning. For instance if one speaks about the attitude of the citizens of South-West London to the Jamaican immigrants in the area one has in mind at one extreme the person—if such a person exists—who would wish to make life impossible and unpleasant for them in every possible way; at the other extreme those who like associating with Jamaicans and who would be prepared to intermarry with them. In constructing a scale to measure attitude to Jamaicans the universe of content from which relevant items may be chosen is that which embraces degrees of 'like' and 'dislike' of Jamaicans between the two extremes mentioned. So there exists a well-defined, though hypothetical, population of items from which a sample may be selected for possible inclusion in the proposed inventory.

Having defined the universe of content the next problem is to assemble a list of items which appear to be a representative sample from the hypothetical population. Possible items are these:

'Are there any Jamaicans amongst your closest friends?'

'Do you feel more uncomfortable than others do if you have to share a seat with a Jamaican on a bus?'

'If you were invited with others to a party run by Jamaicans would you go?'

These items incidentally are representative of three different types of item which occur commonly in attitude inventories. The first is a relatively factual type of item and presents no problem to the respondent. It can be called the *objective* type of item. The second item is an example of the *subjective* type of item. It is not very satisfactory as an answer to it may reflect as much the neuroticism of the respondent as his attitude to Jamaicans. The third item is an example of the *normative* type and is a bit unsatisfactory too in as far as it anticipates some knowledge of the community norms on the part of the respondent and such knowledge may not be available.

Having assembled a number of suitable items the next step is to have them answered by a representative sample of the population whose attitude it is desired to measure. A careful examination of the responses will now reveal any questions which are ambiguous or have been misunderstood by the respondents and these can be deleted. After this initial screening the statistical analysis of the data can begin, the aim being to discover those items which have high discriminative value for differentiating people with strong favourable attitudes from those with strong unfavourable attitudes. The latter will form the first draft of the inventory.

Estimating Item Parameters

After the initial screening of the items a score for each respondent can be obtained from those that remain. Items answered in a manner which indicates a favourable attitude can be scored $+1$, while the remainder are scored zero—all items being of the dichotomous type. If the scores obtained show considerable variation this will be an indication that the items do discriminate between the respondents.

The variation in the scores, too, is indicative of an assumption underlying the procedure of attitude measurement that an attitude continuum (or normal variable) exists on which respondents can be ranked. This variable—for the attitude inventory we have in mind—can be called x. Following the theory of cognitive test construction it is now assumed that the probability, p, that a respondent whose standardised score on the variable x is x' will respond favourably to the j-th item is given by

$$p = \frac{1}{\sqrt{(2\pi)}} \int_{-\infty}^{(x'-\alpha)/\sigma} e^{-u^2/2} \, du \qquad \ldots 11.1$$

where α and σ are constants for that item. When p is $0 \cdot 5$ x' is equal to α, so that—following psychophysical nomenclature—α is called the *limen* of the item, that point on the x-scale below which 50 per cent of the population respond favourably to the item. The constant σ, on the other hand, determines how well the item discriminates between respondents with favourable and unfavourable attitudes. The statistical problem in item analysis is one of estimating the values of α and σ for the different items. Methods of doing this have been given by several writers including Ferguson (1942) and Lawley (1942–3). In addition Finney (1944) has pointed out that the problem of estimating the limina and precisions of test items is analogous to problems in toxicological experiments and he has shown how the methods of probit analysis can be applied to them. More recently it has been recommended that the integrated normal curve in equation 11.1 be replaced by the *logistic* curve (Birnbaum, 1957), and a method of finding maximum likelihood estimates of the limen and precision of test items using this curve has been given by Maxwell (1959).

Unfortunately the labour involved in estimating item parameters by the exact methods just mentioned is great though programmes for doing the calculations on electronic computers will soon be widely available. A good graphical method for use with the logistic function too has been given by Hodges (1958). Meanwhile it is customary to apply some fairly quick, though mathematically less elegant, method in practice. That to be described presently was given in a course of lectures on the construction of cognitive tests by W. G. Emmett at Edinburgh and is inspired by Lawley's paper (1942–3).

Item Analysis

If the proportion of people in the population who answer an item favourably is p then, by the binomial distribution, the variance of that item is pq, where $q = (1-p)$, so that its standard deviation is $\sqrt{(pq)}$. By elementary calculus it can be shown that the variance is a maximum when $p = q = \frac{1}{2}$; it is a minimum when either p or q is zero. It follows that items which are answered favourably by 50 per cent of the population have maximum standard deviations, consequently such items tend to discriminate well between the respondents. That the variance of an item decreases rapidly for p, or q, less than about one-fifth is seen from the following table:

Percentage of favourable responses	p	q	pq
50	$\frac{1}{2}$	$\frac{1}{2}$	0·25
40	$\frac{2}{5}$	$\frac{3}{5}$	0·24
30	$\frac{3}{10}$	$\frac{7}{10}$	0·21
20	$\frac{1}{5}$	$\frac{4}{5}$	0·16
10	$\frac{1}{10}$	$\frac{9}{10}$	0·09

In view of this it is recommendable to omit from the list of items those to which more than 80 per cent or less than 20 per cent of the respondents give favourable answers.

The next essential step in item analysis is to look at the way in which response to each item increases as the total score on the inventory increases. To do this the respondents should be arranged in order of total score and should then be divided into about six equal sub-groups. The number of respondents in each sub-group that answers each item favourably should now be ascertained, and examined. To keep the discussion concrete suppose that the sample size is 300 so that each of six sub-groups contains 50 respondents, and let us consider the hypothetical data given in Table 11.1.

The pattern of scores for the first item is very satisfactory. For Sub-group I only three of the 50 respondents give favourable replies while for Sub-group VI 44 out of 50 do so. Moreover, the number of favourable responses increases linearly from Sub-group I to Sub-group VI. The discriminative value of the item therefore is good and increases in a consistent way from sub-group to sub-group. In addition the proportion of respondents who reply to the item in a favourable manner is exactly 0.5 so the variance of the item is a maximum.

For both Items 2 and 3 the patterns of scores are unsatisfactory, and both items should be rejected. For Item 2, though the number of favourable responses increases linearly, the total proportion of

TABLE 11.1

Number of Respondents out of 50 giving Favourable Responses to Items 1, 2 and 3, in each of Six Sub-groups arranged from Left to Right in order of Increasing Total Score

	Sub-groups						Favourable responses	
Item number	I (50)	II (50)	III (50)	IV (50)	V (50)	VI (50)	Total	Percentage
1	3	12	23	30	38	44	150	50
2	4	6	7	10	12	15	54	9
3	17	36	12	19	28	20	132	44

favourable responses of 0.09 is rather low. For Item 3, though the total proportion of responses (0.44) is very acceptable the relationship between the item and the total score is non-linear. This item then is unsuitable for inclusion in an inventory in which the total score is obtained by a simple additive process of items answered favourably.

In general the question of whether to accept or reject an item will not be as easy to answer as in our three examples. To adopt too strict a criterion of acceptance might lead the experimenter to reject the majority of his original items while new and acceptable ones might be difficult to obtain. If he wishes to have a rule to guide him then, after inspecting for linearity, the data for each item could be arranged in a 2×2 table as shown below and a test of association performed. For example the data for Item 1, when the responses for the three highest and the three lowest groups are combined, are:

	Total score		
Score on item	Low	High	Total
1	38	112	150
0	112	38	150
Total	150	150	300

The value of chi square is $71 \cdot 1$, which is significant at the $0 \cdot 1$ per cent level. Should the experimenter find that he had a proficiency of acceptable items he might wish to reject those for which chi square was not significant at say the 5 per cent level, but if good items were scarce he might have to accept a less stringent level of significance.

Standard errors of inventory scores. When the final selection of items has been made the next step is to obtain some measure of the reliability of the inventory. The measure traditionally used by psychometricians is the *reliability coefficient* from which the 'error of measurement' of individual scores is obtained. But opinions differ as to how best to obtain this coefficient, so it will be advisable to indicate how errors of measurement of inventory scores may be estimated using the methods of analysis of variance.

The items in the inventory are divided randomly into two equal groups: alternatively, if the items can be matched in pairs, one member of each pair can be allotted to each group by tossing a coin. In this way two equivalent inventories, A and B, are obtained. A representative sample is now drawn from the population concerned and it in turn is divided into two equivalent sub-samples.

The two inventories are now administered to the two sub-samples, preferably consecutively so as to keep environmental variables as nearly constant as possible, one sample answering inventory A first the other inventory B first. This experimental design is known as a *cross-over* design. A detailed account of how to analyse data obtained from its use has been given elsewhere (Maxwell, 1958, pp. 82 *et seq.*), but the partitioning of the degrees of freedom are repeated below for the case where there are N respondents and consequently $2N$ scores.

Source of variation	Degrees of freedom
Between occasions of testing (first v. second)	1
Between inventories A and B	1
Between sub-samples	1
Between respondents within sub-samples	$N-2$
Residual	$N-2$
Total	$2N-1$

In the analysis a significant result for the first component would indicate 'practice effect', while one for the fourth would show that the members of the sub-samples were not too homogeneous; but if the randomisation in the experiment is carefully done components two and three should not be significant. The *mean square* for the residual term provides an estimate of the error variance of an individual score on either inventory A or inventory B. An estimate of the error variance of a score obtained on the complete inventory, A plus B, is given by twice the residual mean square—under the assumption that errors for individual A and B scores are uncorrelated. The square root of this quantity, which will be denoted by s, gives an estimate of the error of measurement of an individual score, say X, on the complete inventory. The 95 per cent confidence limits of X are therefore $X \pm 1 \cdot 96s$.

Transforming the scores. When an attitude inventory is composed of items chosen so that the proportion of people responding favourably (or unfavourably) to them is as near as possible to $0 \cdot 5$, the distribution of the scores on the inventory will tend to be rectangular. When this is found to be the case it is customary to transform the obtained distribution into a normal distribution having some suitable mean and standard deviation. When such a transformation is considered to be desirable it should be carried out before an estimate of the 'error of measurement' of individual scores is obtained.

REFERENCES

BIRNBAUM, A. (1957) *Probability and Statistics in Item Analysis and Classification Problems*, U.S.A.F. Report, Randolph Air Force Base, Texas

EDWARDS, A. L. (1957) *Techniques of Attitude Scale Construction*, New York, Appleton

FERGUSON, G. A. (1942) 'Item selection by the constant process', *Psychometrika*, **7**, 19–29

FINNEY, D. J. (1944) 'The application of probit analysis to the results of mental tests', *Psychometrika*, **9**, ?1–9

LAWLEY, D. N. (1942) 'On problems connected with item selection and test construction', *Proceedings of the Royal Society of Edinburgh*, (*A*), Edinburgh, *60*, 64–82.

LORD, F. M. (1953) 'An application of confidence intervals and of maximum likelihood to the estimates of an examinee's ability', *Psychometrika*, **18**, 57–76

MAXWELL, A. E. (1958) *Experimental Design in Psychology and the Medical Sciences*, London, Methuen & Co.

MAXWELL, A. E. (1959) 'Maximum likelihood estimates of item parameters using the logistic function', *Psychometrika*, **24**

FURTHER READING

TORGERSON, W. S. (1958) *Theory and Methods of Scaling*, New York, John Wiley & Sons

APPENDIX

TABLE A

Percentage Points of the χ^2 Distribution

D.F.	Probability (P)			
	0·050	0·025	0·010	0·001
1	3·841	5·024	6·635	10·828
2	5·991	7·378	9·210	13·816
3	7·815	9·348	11·345	16·266
4	9·488	11·143	13·277	18·467
5	11·071	12·833	15·086	20·515
6	12·592	14·449	16·812	22·458
7	14·067	16·013	18·475	24·322
8	15·507	17·535	20·090	26·125
9	16·919	19·023	21·666	27·877
10	18·307	20·483	23·209	29·588
11	19·675	21·920	24·725	31·264
12	21·026	23·337	26·217	32·909
13	22·362	24·736	27·688	34·528
14	23·685	26·119	29·141	36·123
15	24·996	27·488	30·578	37·697

D.F.	0·050	0·025	0·010	0·001
16	26·296	28·845	32·000	39·252
17	27·587	30·191	33·409	40·790
18	28·869	31·526	34·805	42·312
19	30·144	32·852	36·191	43·820
20	31·410	34·170	37·566	45·315
21	32·671	35·479	38·932	46·797
22	33·924	36·781	40·289	48·268
23	35·173	38·076	41·638	49·728
24	36·415	39·364	42·980	51·179
25	37·653	40·647	44·314	52·620
26	38·885	41·923	45·642	54·052
27	40·113	43·194	46·963	55·476
28	41·337	44·461	48·278	56·892
29	42·557	45·722	49·588	58·302
30	43·773	46·979	50·892	59·703
40	55·759	59·342	63·691	73·402
50	67·505	71·420	76·154	86·661
60	79·082	83·298	88·379	99·607
80	101·879	106·629	112·329	124·839
100	124·342	129·561	135·807	149·449

Index

161